AF595471

Emerging and Disruptive Next-Generation Technologies for POC: Sensors, Chemistry and Microfluidics for Diagnostics

Emerging and Disruptive Next-Generation Technologies for POC: Sensors, Chemistry and Microfluidics for Diagnostics

Editors

Maria Serena Chiriacò
Elisabetta Primiceri
Francesco Ferrara

MDPI • Basel • Beijing • Wuhan • Barcelona • Belgrade • Manchester • Tokyo • Cluj • Tianjin

Editors

Maria Serena Chiriacò
Institute of Nanotechnology of CNR (CNR NANOTEC)
Consiglio Nazionale delle Ricerche (CNR)
Lecce
Italy

Elisabetta Primiceri
Institute of Nanotechnology of CNR (CNR NANOTEC)
Consiglio Nazionale delle Ricerche
Lecce
Italy

Francesco Ferrara
STMicroelectronics - Sede di Lecce
STMicroelectronics
Lecce
Italy

Editorial Office
MDPI
St. Alban-Anlage 66
4052 Basel, Switzerland

This is a reprint of articles from the Special Issue published online in the open access journal *Micromachines* (ISSN 2072-666X) (available at: www.mdpi.com/journal/micromachines/special_issues/Technologies_POC).

For citation purposes, cite each article independently as indicated on the article page online and as indicated below:

LastName, A.A.; LastName, B.B.; LastName, C.C. Article Title. *Journal Name* **Year**, *Volume Number*, Page Range.

ISBN 978-3-0365-3416-9 (Hbk)
ISBN 978-3-0365-3415-2 (PDF)

Contents

About the Editors

Maria Serena Chiriacò

Maria Serena Chiriacò gained her master's degree cum laude in Human Biology in 2008 at Università del Salento; from May 2009 to June 2012, she attained her PhD on the "Intelligent Systems and Technologies"course at Università del Salento, Scuola Superiore ISUFI, Lecce, Italy, at the end of which she discussed her PhD thesis entitled "Protein EIS biosensors for on-chip diagnostics". Currently, she works as a permanent researcher at Nanotec Lecce. She gained high-level experience in the field of Lab-On-Chip devices for on-field diagnostics. Her skills vary from the design, fabrication and applications of electrochemical sensors to microfluidics and on-chip pre-treatments of biological samples, as confirmed by the publication of valuable papers in the field. The biological applications of the developed microfluidic and sensing platforms range from diagnostic tools to study biorecognition events (antigen/antibody, cell–cell, antibodies/cells interactions) to the detection of allergens or toxins from food or environmental samples, to the on-chip separation of particles. She also has expertise in optical, laser and soft lithography and in employing photosensitive polymeric materials for microfluidic applications.

Elisabetta Primiceri

Elisabetta Primiceri (1982) received her master's degree (magna cum laude) in Industrial and Molecular Biotechnology in 2006 at University of Bologna and in 2011 she finished her PhD on "Interdisciplinary Science and Technology"at University of Salento, Lecce, Italy, with her thesis entitled "Cell-Chip: new tools for cell biology". Since 2011, she has worked as a post-doc researcher at Consiglio Nazionale delle Ricerche (CNR) in Lecce and from January 2019 she has been a permanent researcher at CNR Institute of Nanotechnology (CNR-Nanotec). Her activities are focused on the development of biosensors (especially electrochemical and plasmonic detection) and new materials for sensing applications. She is also focused on the development of sensing platforms for cell biology. Her skills include microfabrication techniques (photolithography and soft lithography), electrochemical measurements, and the electrosynthesis of polymers. She has also gained experience in cell culture and cell biology.

Francesco Ferrara

Francesco Ferrara is an application engineer for the System Research & Applications (SRA) group of STMicroelectronics. He holds an MSc in microelectronics engineering from Polytechnic University of Bari (IT) and a PhD in biomolecular nanotechnology from University of Salento (IT). From 2004 to 2010, he worked on the first diagnostic platform based on a disposable silicon chip for molecular testing in an ST microfluidic business unit. Since 2010, he has been a part of Silicon Biotech group and he works on new disposable Real-Time PCR platforms for molecular diagnosis with integrated sample preparation devices. He is an expert in cleanroom fabrication, optoelectronics and microfluidics, the co-PI of the Attract SMILE project and involved in national and European funding projects in the field of new generation technologies for healthcare.

Editorial

Editorial for the Special Issue on Emerging and Disruptive Next-Generation Technologies for POC: Sensors, Chemistry and Microfluidics for Diagnostics

Francesco Ferrara [1,2,*], **Elisabetta Primiceri** [2,*] and **Maria Serena Chiriacò** [2,*]

1 STMicroelectronics s.r.l., Via per Monteroni, 73100 Lecce, Italy
2 CNR NANOTEC Institute of Nanotechnology, Via per Monteroni, 73100 Lecce, Italy
* Correspondence: francesco.ferrara@st.com (F.F.); elisabetta.primiceri@nanotec.cnr.it (E.P.); mariaserena.chiriaco@nanotec.cnr.it (M.S.C.)

Recently, the attention paid to self-care tests and the need for easy and large-scale screenings of a high number of people has dramatically increased. The COVID-19 pandemic has emphasized the requirement for affordable tools for the safe management of biological fluids and distanced diagnostic procedures. Limiting the diffusion of infections has emerged as a compulsory requirement, especially to lighten the pressure on public healthcare institutions. Obviously, other kinds of pathologies (cancer or other degenerative diseases) continue to call for attention for earlier and more widespread diagnoses and treatments.

In this scenario, the research field of point-of-care (POC) diagnostics could strongly contribute to the realization of a valid alternative to standard tools, in order to hold off the spread and aggressiveness of the pandemic whilst not neglecting the normal activity of prevention and care related to other diseases. Indeed, in the last two years, cancer (just as an example) has seen a significant drop in newly diagnosed cases and the management and follow-up of active cases has been made difficult by the limited access to healthcare institutions and by the postponement of programmed visits and interventions. The same limits have also hindered the monitoring of other types of pathologies which require continuous care, such as metabolic disorders or prenatal assays, with a significant impact on common health, society and economy. That is why the possibility of having available instruments for the self-collection of specimens, distanced diagnostics and telehealth tools could strongly help in facing the current pandemic assault, but also finally give a chance to all the smart technologies developed during the last 10–15 years to overcome that gap between academia and the market which has kept them away from current practice.

The contribution to the development of this research field comes from the areas of innovative plastic and 3D microfluidics, smart chemistry and the integration of miniaturized sensors, going in a direction to improve the performances of in vitro diagnostic (IVD) devices and addressing the new challenges requiring patients' compliance and minimal interactions with both medical personnel and clinical settings. On the other hand, the need for self-contained tools to be wireless, programmable and able to continuously monitor the state of health or illness of an individual allowed wearable technology to make huge strides forward. Moreover, the key enabling technologies developed can be useful to facilitate not only clinicians but also researchers, improving reproducibility, saving time and money for reagents and avoiding tests on animals for technologies related to organ-on-chip or tissue engineering.

In our Special Issue, we collected papers describing easy strategies to identify diseases at the point-of-care level, but also dealing with innovative biomarkers, sample treatments and chemistry and engineering advances which, in perspective, represent promising tools to be applied to the field. This Special Issue mainly stems from the EDGE-Tech (Emerging and Disruptive next-Generation Technologies for POC) Workshop organized in the frame of the

Citation: Ferrara, F.; Primiceri, E.; Chiriacò, M.S. Editorial for the Special Issue on Emerging and Disruptive Next-Generation Technologies for POC: Sensors, Chemistry and Microfluidics for Diagnostics. *Micromachines* **2022**, *13*, 181. https://doi.org/10.3390/mi13020181

Received: 25 January 2022
Accepted: 25 January 2022
Published: 26 January 2022

EU SMILE-Attract project, devoted to the development of innovative strategies for the early diagnostics of oral cancer from saliva. The work of Zoupanou et al. [1] describes the SMILE platform and its microfluidic approach to the detection of cancer cells into plastic and low-cost microchannels. In particular, plug-n-play tools for microfluidics have been obtained in PMMA, a versatile material that is low-cost and easy to handle, overcoming the limits of PDMS-based methods which are less robust and not suitable for industrial exploitation. The 3D network of microchannels and the complete transparency of PMMA, even after micromilling and bonding, allows for the realization of buried paths and a contemporary check on the whole process happening at the different levels. Moreover, the conjunction of channels at the interconnection point can be used to evaluate the mixing or gradient generation in the final portion of common channels. The final goal of the SMILE Project is the complete on-chip sample handling and analysis, allowing the addition of reagents to specimens, their mixing and the detection of searched analytes into a specific area of the device. In this case, a serpentine path was functionalized with the immobilization of antibodies against the membrane antigen EpCAM, which is able to distinguish cancer cells from blood cells.

In order to use microfluidics to operate label-free and affordable cell analysis, Professor Zattoni and co-workers developed a system named "Celector®", which is able to analyze, discriminate and separate a wide range of cell mixtures based on their physical characteristics with high resolution and throughput. By implementing Non-Equilibrium Earth Gravity Assisted Dynamic fractionation (NEEGA-DF), the system uses a micro-camera for cell detection. The developed instrument makes use of specifically designed software for image acquisition, post-processing and data analysis, obtaining a multiparametric fractogram representing the number, size and shape of the eluted cells as a function of fractionation time. This results in a complete fingerprint of the cell sample. In the paper proposed for this Special Issue, this technology is applied to perform the quality check of freshly isolated amniotic epithelial cells AECs [2]. Comparing the possible differences in cells' yield and composition of amniotic membrane, the live fractogram was used as a predictive model to successfully define the isolation procedure. Post-processing image data were compared to biological data of cell recovery, cell vitality and adhesion ability, identifying a new predictive tool for laboratories and cell banks that isolate and cryopreserve fetal annex stem cells for research and future clinical applications.

The possibility to continuously check the state of health or illness of an individual in a non-invasive or minimally invasive manner has seen a groundbreaking input from wearable technologies. In this regard, the constant monitoring of body parameters such as pressure, heart rate or electrolyte concentration could significantly help in the prevention of cardiovascular diseases or other metabolic impairments, gaining high compliance from patients and encouraging a large audience of final users to screenings. With this aim, the group of Professor De Vittorio contributed to our Special Issue with a paper describing the realization of a flexible glucose sensor suitable for sweat analysis, paving the way for a new generation of non-invasive glucose sensors and improving the quality of life of diabetic patients [3]. In particular, they obtained a three-electrode device by the thermal evaporation of gold or silver (for working/counter and reference electrodes, respectively) on a polystyrene foil. Moreover, a versatile nanoimprinting process for microfluidics was available to ease sampling. For the sensing layer, a gold electrode was modified with a cysteine layer and glutaraldehyde cross-linker for enzyme probe immobilization. To demonstrate the reliability of their glucose sensor, chronoamperometric measurements were performed in a PBS-buffered glucose solution in a linear range between 0.025 mM and 2 mM.

Following the direction of key enabling technologies able to foster the development of research practice and applications, our Special Issue also hosted two reviews with different topics but both dealing with innovative solutions encompassing the field of chemistry and engineering.

In the work of Cinquino et al., the huge diffusion of fabric with augmented functionalities, enabling the integration of displays, sensors and other electronic components into textiles, is discussed [4]. Typical examples of wearable and portable devices are smart watches, glasses, wristbands and belts which can be easily applied to monitor health status or exercise parameters. These kinds of devices are usually made of rigid and planar materials, making them uncomfortable to wear. In this review paper, the possibility to integrate microelectronic devices with fabrics is deepened as an important innovation toward the development of a more comfortable and versatile technology. In particular, the authors considered light-emitting diodes (LEDs), alone or coupled with polymer optical fibers (POFs), as the most robust technology. Moreover, OLEDs (Organic LEDs) were also addressed as a very promising approach for the future of light-emitting fabrics, even though some issues still need to be resolved.

The other review published in the Special Issue deals with tissue engineering. The possibility to produce the in vitro/in vivo regeneration of tissue by tuning the characteristics of the extracellular matrix or the fate of stem cells is one of the most fascinating technologies evolved in the last decades. The development of tissue engineering has gone through advances in organ-on-chip technologies which have given a strong input toward the use of individual cells to re-create conditions in which tissue development and cell differentiation mechanisms could be followed or drugs could be tested without recurring to animal models. The challenge research has to face in the field of tissue engineering is mainly related to the real and long-lasting biocompatibility of the scaffold and materials used, but the results are potentially revolutionary. In their work, Mijanović, Pylaev et al. discussed the capabilities and limitations of biomaterial scaffolds to successfully integrate into the surrounding environment in the restoration mechanisms of damaged tissues. In particular, biocompatibility with host tissues remains crucial for preventing infection and promoting implant integration. The molecular and cellular modulation indeed is important and necessary for successful graft integration, post-surgery clinical outcome and long-term survival. The authors specifically addressed corneal tissue engineering, which has attracted great interest recently, due to the attempts to avoid many of the complications encountered in traditional donor corneal transplantation [5]. The involvement of cells could not only allow the creation of a cornea analog, which should provide high optical transparency, mechanical integrity and proper cell behavior, leading to effective re-epithelialization, but also give promises for the full regeneration and integration of the graft.

As Guest Editors of "*Emerging and Disruptive Next-Generation Technologies for POC: Sensors, Chemistry and Microfluidics for Diagnostics*", we would like to take this opportunity to thank all the authors for submitting their papers to this Special Issue. We would also like to acknowledge all the reviewers for their time dedicated to suggesting improvements to the quality of the submitted papers and of the whole Special Issue.

Conflicts of Interest: The authors declare no conflict of interest.

References

1. Zoupanou, S.; Volpe, A.; Primiceri, E.; Gaudiuso, C.; Ancona, A.; Ferrara, F.; Chiriacò, M.S. SMILE Platform: An Innovative Microfluidic Approach for On-Chip Sample Manipulation and Analysis in Oral Cancer Diagnosis. *Micromachines* **2021**, *12*, 885. [CrossRef] [PubMed]
2. Zia, S.; Martini, G.; Pizzuti, V.; Maggio, A.; Simonazzi, G.; Reschiglian, P.; Bonsi, L.; Alviano, F.; Roda, B.; Zattoni, A. A New Predictive Technology for Perinatal Stem Cell Isolation Suited for Cell Therapy Approaches. *Micromachines* **2021**, *12*, 782. [CrossRef] [PubMed]
3. Müsse, A.; La Malfa, F.; Brunetti, V.; Rizzi, F.; De Vittorio, M. Flexible Enzymatic Glucose Electrochemical Sensor Based on Polystyrene-Gold Electrodes. *Micromachines* **2021**, *12*, 805. [CrossRef] [PubMed]
4. Cinquino, M.; Prontera, C.T.; Pugliese, M.; Giannuzzi, R.; Taurino, D.; Gigli, G.; Maiorano, V. Light-Emitting Textiles: Device Architectures, Working Principles, and Applications. *Micromachines* **2021**, *12*, 652. [CrossRef] [PubMed]
5. Mijanović, O.; Pylaev, T.; Nikitkina, A.; Artyukhova, M.; Branković, A.; Peshkova, M.; Bikmulina, P.; Turk, B.; Bolevich, S.; Avetisov, S.; et al. Tissue Engineering Meets Nanotechnology: Molecular Mechanism Modulations in Cornea Regeneration. *Micromachines* **2021**, *12*, 1336. [CrossRef] [PubMed]

micromachines

MDPI

Article

SMILE Platform: An Innovative Microfluidic Approach for On-Chip Sample Manipulation and Analysis in Oral Cancer Diagnosis

Sofia Zoupanou [1,2], Annalisa Volpe [3,4], Elisabetta Primiceri [2], Caterina Gaudiuso [3,4], Antonio Ancona [3,4], Francesco Ferrara [2,5,*,†] and Maria Serena Chiriacò [2,*,†]

1 Department of Mathematics & Physics E. de Giorgi, University of Salento, Via Arnesano, 73100 Lecce, Italy; sophia.zoupanou@nanotec.cnr.it
2 CNR NANOTEC—Institute of Nanotechnology, Via per Monteroni, 73100 Lecce, Italy; elisabetta.primiceri@nanotec.cnr.it
3 Physics Department, University of Bari "Aldo Moro", Via Orabona 4, 70126 Bari, Italy; annalisa.volpe@uniba.it (A.V.); caterina.gaudiuso@uniba.it (C.G.); antonio.ancona@uniba.it (A.A.)
4 Institute for Photonics and Nanotechnologies (IFN), National Council of Research of Italy (CNR), 70126 Bari, Italy
5 STMicroelectronics s.r.l., Via per Monteroni, 73100 Lecce, Italy
* Correspondence: francesco.ferrara@st.com (F.F.); mariaserena.chiriaco@nanotec.cnr.it (M.S.C.)
† The two authors contributed equally to the work.

Citation: Zoupanou, S.; Volpe, A.; Primiceri, E.; Gaudiuso, C.; Ancona, A.; Ferrara, F.; Chiriacò, M.S. SMILE Platform: An Innovative Microfluidic Approach for On-Chip Sample Manipulation and Analysis in Oral Cancer Diagnosis. *Micromachines* **2021**, *12*, 885. https://doi.org/10.3390/mi12080885

Academic Editor: Nam-Trung Nguyen

Received: 16 July 2021
Accepted: 26 July 2021
Published: 27 July 2021

Publisher's Note: MDPI stays neutral with regard to jurisdictional claims in published maps and institutional affiliations.

Abstract: Oral cancer belongs to the group of head and neck cancers, and, despite its large diffusion, it suffers from low consideration in terms of prevention and early diagnosis. The main objective of the SMILE platform is the development of a low-cost device for oral cancer early screening with features of high sensitivity, specificity, and ease of use, with the aim of reaching a large audience of possible users and realizing real prevention of the disease. To achieve this goal, we realized two microfluidic devices exploiting low-cost materials and processes. They can be used in combination or alone to obtain on-chip sample preparation and/or detection of circulating tumor cells, selected as biomarkers of oral cancer. The realized devices are completely transparent with plug-and-play features, obtained thanks to a highly customized architecture which enables users to easily use them, with potential for a common use among physicians or dentists with minimal training.

Keywords: oral cancer; circulating tumor cells; micromixers; 3D microfluidics; biodetection; plastic microfluidics; microfabrication

1. Introduction

Head and neck cancers represent the sixth most common type of cancer in Europe, accounting for 150,000 new patients per year, and 60% of patients with advanced disease at diagnosis die within 5 years [1,2]. The most widespread cancer in the head and neck region is oral squamous cell carcinoma (OSCC), occurring at the border of lips and/or at the posterior of the tongue or palates [3]. Many factors can increase the probability of the disease; tobacco and alcohol consumption [4], oncogenic viruses (e.g., papillomavirus, HPV, or Epstein–Barr virus EBV) [5] and poor oral health [6] are known risk factors. Genetic predisposition is also a key consideration when studying the development of oral cancer. As an example, the role of *NFKB1* gene polymorphisms are currently under investigation [7]. The link of germline genetics and environmental factors to pathologic phenotypes can contribute to a better understanding of the interactive role of the environment, tumor cells, immune cells and microbiome in various diseases [8,9]. Moreover, the importance of correct diet and lifestyle is another crucial aspect in preventing oral cancer, as they also modulate the oral microbiome, which has been demonstrated to play a role in cancer onset, particularly due to its influence in the modulation of immune system [10]. High levels of colonization of OSCC by facultative oral streptococci have been shown [11], and, more recently, Zhang and coworkers compared the microbiota compositions of tumor sites and normal tissues, finding bacteria significantly associated with mouth tumors [12]. These aspects fall in the gene-by-environment (G $\times$ E) interaction range, which considers a wider analytical method to study the onset of diseases, including the integration of microbiology into molecular pathology and epidemiology models, going in the direction of more personalized medicine [9].

Currently, biomarkers of oral cancer are indeed inadequate, and inflammatory molecules may have a low specificity. Moreover, the detection of precancerous lesions is not routinely carried out in clinical settings [13]. An easy-to-use, noninvasive assay is strongly needed, and the possibility to perform tests on saliva is an attractive strategy to increase patient compliance [14].

In order to improve the quality of diagnostic and prognostic early screening tests, as well as their availability for a large cohort of potential patients, a number of biomarkers from body fluids have been identified. Among these, inflammatory biomarkers [15–17] and circulating tumor cells are the most promising entities to be found in blood or saliva, even at the early stage of the disease, and this method of detection has the potential to be translated into on-chip platforms. Early detection of OSCC is indeed the only way to limit the consequences of disease, and the main challenge in prevention is large-scale screening [18], together with an appropriate diet and correct lifestyle [4]. To reach this goal, the need for a noninvasive assay is compulsory, and the possibility to perform tests on saliva is an attractive strategy to make this test suitable for all, in addition to the ambitious objective of the SMILE platform.

The main purpose of SMILE project is the development of a low-cost sensing device for oral squamous cell carcinoma (OSCC) early screening, with features of high sensitivity, portability, and ease of use, with the aim of reaching a large audience. OSCC is usually diagnosed at an advanced stage, where highly invasive surgery and chemotherapy are required, heavily compromising life quality and survival [3,19].

Breakthrough technologies of the platform include the optimization and integration of a highly customizable plastic microfluidics device able to perform some of the most common operating tools while handling samples, particularly micromixing, gradient generation [20], and the capture and detection of small objects such as circulating tumor cells [21].

Moreover, the recent spread of COVID-19 and its huge impact on clinical settings has forced healthcare systems to undergo a total rearrangement of rules and priorities. In cancer management, this has led to weighing up the risks of tumor progression due to a delay in treatments against the potential of adding to the hospital burden by increasing the risk of exposure to SARS-CoV2. The need of new tools able to maintain standards of diagnosis and control of active tumor cases, while limiting infections, through rapid sample collection is then compulsory. New instruments, based on a lab-on-chip (LoC) approach, with features of low cost and a plug-and-play setup, along with the aim of using body fluids easily collected without the need for healthcare personnel, could revolutionize the approach to periodic screening and follow-ups, allowing the possibility to perform them in a safe and distanced manner.

To meet this goal, microfluidic technologies and the use of plastic substrates in combination with rapid prototyping methods, i.e., fs laser technology and micromilling, seem to be a good alternative to standard methods in the realization of polymeric lab-on-chip, without any constraints on the substrate material [22,23].

Today, several techniques can be exploited for the rapid prototyping of polymeric LoC [24,25]. In particular, thanks to its high resolution (<50 nm), soft lithography is one of the most exploited methods for the rapid prototyping of a polymeric microfluidic device [26,27]. This technology requires the fabrication of a mold, typically by photolithography, as well as replication and assembly of the entire device. Consequently, like other similar molding techniques [28], soft lithography is a time-consuming process, which often limits optimization in prototyping or iterative design. Thus, despite its very low cost, it does not allow an easy and direct translation into large-scale production and industrial exploitation, with the aim of reaching the market of in vitro diagnostics (IVD). A further limitation of soft lithography is the material used. Polydimethylsiloxane (PDMS) microchannels subjected to high liquid pressure undergo deformation [29].

Technologies based on the direct microstructuring of the substrate and, thus, not requiring a mold have been proven to be more suitable during the design of a new device. Among such technologies, three-dimensional (3D) printing gives the possibility of fabricating low-cost 3D microfluidic devices in a single step from a computer model [30,31]. The major concerns about this technique regard the inability to reliably print microfluidic channels with dimensions less than several hundred microns, dimensional fidelity, surface quality, optical transparency, and reduced choice of materials [32]. Conversely, ultrafast laser technology, as a non-clean room process, provides a convenient, economical, and flexible way to fabricate micrometric fluidic patterns by varying the laser parameters [33,34], avoiding the expensive and time-consuming production of masks. The fs laser enables a "cold" ablation of the irradiated volume, which allows the material to be removed by ablation from the irradiated area with negligible thermal damage to the surrounding substrate [35], ensuring high precision and up to submicrometric resolution [36]. Moreover, fs laser pulses do not pose any restriction on the substrate materials [37]. However, the principal constraints of this technique are the high costs of the laser source and them not being efficiently suitable for the fabrication of over-micrometric structures.

Mechanical micromilling is a flexible, cost-efficient, rapid prototyping technology for polymer device machining. In comparison with fs laser, it is more convenient for large features [38]. However, it results in poor surface quality and resolution [39].

The possibility to use plastic devices will significantly improve the reproducibility and stability of experimental setups compared with PDMS-based state-of-art microfluidics. PDMS, despite its features of low cost and disadvantages due to the presence of a hard master obtained by lithographic methods, often suffers from sealing leakage, poor connection stability, consequently low reproducibility of experiments. This aspect is not secondary when dealing with biological methods, as complexity can be achieved only by avoiding variability of the boundary conditions given, i.e., by the device.

Moreover, the possibility to obtain stable connections through customized inlet and outlet holes allows the ease of use necessary to achieve so-called "world-to-chip" connections [40], avoiding the use of magnetic gaskets, clamps, or glue which are usually expensive and time-consuming, and which do not allow reusing the device or capillary tubes [41].

On the other hand, traditional fabrication methods such as lithographical techniques, used to create micro and nanoscale structures, are very useful for prototyping and research experiments, but they do not translate well into mass production, in addition to the use of PDMS, thus limiting its commercial applications owing to the difficulty in upscaling manufacturing and the relatively high cost compared to polymeric alternatives [42]. More critically, intrinsic material properties of native PDMS represent another chapter of drawbacks, such as evaporation, leaching, and absorption of a flowed liquid sample which often makes PDMS unsuitable for repeatable, robust microfluidic biological and chemical analysis applications [43]. Postprocessing of the material (for example, parylene coating) can overcome these limitations but adds an additional backend processing stage and, therefore, makes it more undesirable for commercial manufacturing [44]. Although glass could be used as an alternative, thermoplastics (plastics) are better from a cost and fabrication perspective, allow easy surface treatment, and are generally transparent and biocompatible [45].

The combination of femtosecond laser-based micromanufacturing and micromilling technologies, associated with robust, disposable, plastic substrates and an innovative sealing method and dedicated surface chemical treatment of microchannels for their biological functionalization, has been recently demonstrated for the development of microfluidic devices for biomedical applications. In particular, it has been shown that combining the micrometric precision offered by ultrashort pulsed laser ablation with the higher machining rate of mechanical micromilling is very beneficial for the rapid and flexible prototyping of polymeric lab-on-chips [46]. The as-fabricated devices can be exploited in applications ranging from the simple on-chip study of cells to the onsite and early diagnosis of diseases [47].

The degree of efficiency revealed by the hybrid microfabrication platform proposed in this work would allow producing 3D microfluidic devices, while embedding additional functionalities such as micromixers and gradient generators. This would push the on-chip platform toward the sample in/answer out concept of point-of-care devices.

In the frame of SMILE (SAW-MIP Integrated Device for Oral Cancer Early Detection) project, we explored several aspects of technology, spanning from simulation of the entire platform through finite element methods (FEM) to tests with artificial samples, as well as from a simulation with particle mixing and a gradient generator to the use of microfluidics using oral cancer cells, envisioning the entire on-chip manipulation and analysis of the sample, with a *plug-and-play* device. In this study, two devices, a micromixer developed on two-level microchannels and a serpentine pathway for the biorecognition of circulating tumor cells, which could work separately or in a subsequential manner, were described. The two devices constitute two building blocks toward the realization of a platform including both sample preparation and biodetection. We performed a completely on-chip functionalization of PMMA, demonstrating a step forward with respect to the current literature, as we obtained the desired results without using complex methods such as UV-curable functionalization or low-pressure radiofrequency (RF) air plasma [48,49]. Anti-EpCAM antibody was chosen as a capture probe to distinguish cancer from noncancer cells in a mixture obtained by keeping together oral cancer-derived cells with blood-derived cells [50], thus mimicking the presence of circulating tumor cells (CTCs) in blood samples. In principle, this antibody can be replaced with any other to identify different kind of cells or extracellular vesicles (microvesicles/exosomes), in order to obtain a liquid biopsy from other biological fluids such as saliva or urine. The investigation of CTCs as early biomarkers of cancer is one of the most promising topics in liquid biopsy, and the translation of this research into a point-of-care device is being explored for the high value which an easy-to-use tool could bring in early diagnosis good practices. To this aim, many microfluidic devices for the separation of CTCs and sample enrichment have been realized, including microsieve integration [51–53], inertial microfluidics [34], serpentine path, and many other chip architectures. However, in most reported cases, microfabricated microsieves are assembled into PDMS-based lab-on-chips, with the inclusion of a membrane, resulting in a low-exploitable approach. In the case of inertial microfluidics, although demonstrated to be very effective in separation, the technique requires a complex simulation and experimental validation phase [54] to define the right geometry as a function of the properties (e.g., dimensions, Young's modulus) of the cells to be separated. As the aim of our work was the development of a low-cost easy-to-use device, with the goal of reaching a large audience of possible users and realizing real prevention of the disease, we preferred to use a simple and common serpentine microchannel, in order to lengthen the path that cells are forced to run, with the aim of maximizing the possibility to be captured by immobilized antibodies.

2. Materials and Methods

2.1. Materials

Both devices were fabricated by assembling different squared layers of transparent PMMA (Vistacryl CQ; Vista Optics, Gorsey Lane, Widnes, Cheshire, UK). Each layer was machined differently according to the design of the device. For the micromixer device, three 30×30 mm^2 PMMA layers were used. The bottom and intermediate layers hosting the micromilled channels were 1 mm thick, while the upper layer containing the inlet/outlet holes was 2.5 mm thick. For the cancer cell capturing device, two square 25×25 mm^2 layers were used. The top layer was 5 mm thick, while the bottom one was 1 mm thick.

The bonding between layers was performed with pure isopropyl alcohol (Sigma-Aldrich, St. Louis, MO, USA).

Surface functionalization of the microchannels was required, following a different procedure for each device. Particle mixing required surface passivation, which included O_2 plasma surface treatment and incubation with 1 mg/mL bovine serum albumin (BSA) (1%) in phosphate-buffered saline (PBS) buffer (Sigma-Aldrich, St. Louis, MO, USA).

The cell capturing, instead, initially included the usage of 3-aminopropyltrirthoxysilane (APTES 5%) in ethanol, glutaraldehyde (0.05%) in water, bovine serum album (BSA) (1%), and Tween®-20 (0.05%) in phosphate-buffered saline (PBS). In addition, we used EpCAM mouse monoclonal antibodies (all reagents from Sigma-Aldrich, USA). It is worth emphasizing that the anti-EpCAM antibodies have the advantage of not being reactive with normal or neoplastic nonepithelial cells and recognize only human EpCAM expressed on the surface of the epithelial cells. A secondary labeled antibody anti-mouse IgG (whole molecule)–FITC antibody produced in goat (Sigma-Aldrich, USA) was used for the fluorescence confirmation assay.

The sealing and working principle of both devices was evaluated by performing a series of tests. For the sample injection/pumping and flow control, we used the Elveflow microfluidic setup (Elvesys, Paris, France), suitable for finely tuning flow injection in a range of 0.4–7 μL/min. For real-time acquisition, we used an Axio Zoom V16 fluorescence microscope (Zeiss, Oberkochen Germany), with an ApoZ1x objective and a numerical aperture (NA) of 0.25.

For the micromixers, we carried out two validation tests using colored inks and fluorescent polystyrene microspheres, with diameters of 200 nm (green) and 1 μm (red) (FluoSpheres ® Fluorescent Microspheres, Invitrogen, Ltd. 3 Fountain Drive Inchinnan Business Park, Paisley, UK) in ethanol (Sigma-Aldrich, St. Louis, MO, USA). Instead, for the capturing of tumor cells, we injected cells derived from the OECM-1 human oral squamous carcinoma cell line (purchased from SCC/Sigma-Aldrich) and Jurkatt cell line (leukemic T-cell lymphoblast from ATCC).

2.2. Computational Modeling

The micromixing tool was simulated by modeling a 3D h-junction in Comsol Multiphysics 5 (COMSOL, Inc., Burlington, MA, USA), using the Microfluidics module CFD package and, specifically, "mixture model, laminar flow" for predicting the fluid flow and particle transport.

Navier–Stokes equations were used to predict the fluid flow through the 3D network of interconnected channels.

As it can be seen, we established two different microchannels with two independent inlets, sharing a common outlet. As boundary conditions, we set an equal uniform velocity/flow rate on both inlets, and zero pressure was applied at the outlet. Furthermore, from the material library, we selected pure water as the defined fluid to cover all domains inside the network of the channels, and we applied zero-flow conditions at the channel walls. The fluid temperature in the entire simulation was set to 25 °C for water. The fluid was considered incompressible, Newtonian, and with no gravitational effects anywhere in the device. Once the fluid flow regime was parameterized, set and tested, we created a surrogate model for particle mixing. Particle inlets and outlet were arranged following the same logic as the fluidics. Moreover, the applied particle parameters were those of the standard polystyrene beads, using two different sizes of 200 nm and 1 μm, respectively. The drag force was set in accordance with the material's properties. The type of mesh built for our geometry was free tetrahedral. These conditions were kept constant throughout the simulation. The constructing response was tested for evaluation, after setting all the necessary parameters, by checking the simulation results for the distribution of the flow, velocity, pressure, and particle distribution at all crucial domains of the design.

Selecting the proper channel design with the optimal mixing performance was one of the main steps in optimization. To this end, to prove the quality of our selected model, we examined the flow behavior when experimented with a range of shapes for one of the microchannels, including rectangular, rhomboidal, and elliptical designs for the mixing chamber. The numerical model used was validated by implementing a real system.

2.3. Design, Fabrication, and Sealing of PMMA Substrates

In order to fabricate the microchannels for the mixing module, with desired dimensions of 200 µm width and 200 µm height, and with specific holes as inlet and outlets, we utilized PMMA substrates and the Mini-Mill/GX micromilling machine (Minitech Machinery, Norcross, GA, USA) with a 200 µm two-flute carbide micro end milling tool. The microfluidic network was designed using Solidworks CAD software (SolidWorks Corporation, 300 Baker Avenue, Concord, MA, USA) and transported in machine code file for micromilling control through computer-aided manufacturing (CAM) software. A 150 mm/min feed rate was used to mill the PMMA layers at 20,000 rpm. The alignment of the inlet and outlet with the channels was achieved using an on-board camera of the micromilling machine. The geometry of the micromixer and the aspect of the final assembled device are shown in Figure 1a–d.

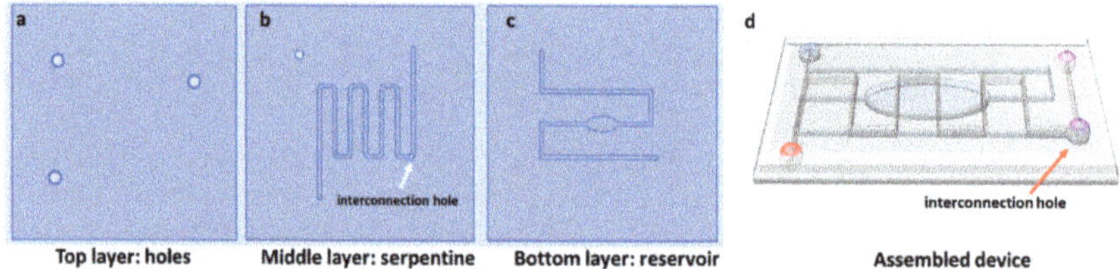

Figure 1. Design of the three layers (**a**) top; (**b**) serpentine; (**c**) bottom and aspects of the assembled device (**d**).

The serpentine channel used for the cancer cell capturing experiments was fabricated by exploiting the femtosecond laser milling process, as previously described in [21]. We used an ultrafast solid-state laser system (mod. TruMicro Femto Ed.; TRUMPF GmbH+ Co. KG, Ditzingen, Germany) based on the chirped pulse amplification technique, which delivers linearly polarized 900 fs pulses at a wavelength of 1030 nm with an almost diffraction limited beam (M2~1.3). The laser beam was circularly polarized by a quarter-wave plate and then focused and moved onto the target surface through a galvo-scan head (IntelliSCANNse 14; SCAN-LAB, Puchheim, Germany) equipped with a telecentric lens of 100 mm focal length. The spot diameter at the focal plane was about 25 µm. The fs laser milling process was carried out by removing the material layer by layer, superimposing two perpendicular scanning paths.

The working parameters used for the serpentine channel fabrication are reported in Table 1.

Table 1. Laser micromilling parameters.

Repetition Rate (RR)	Pulse Energy	Scan Speed	Hatch Distance
50 kHz	12 µJ	40 mm·s^{-1}	5 µm

After the fs laser process, loosely attached debris was removed by ultrasonic cleaning in distilled water for 10 min. The dimensions of the fs laser-milled microfeatures were measured using an optical microscope (Nikon Eclipse ME600). Moreover, the average roughness Ra of the milled surface was measured by means of an optical ContourGT InMotion (Bruker, Billerica, MA, USA) profilometer with nanometric resolution and was estimated to be <2 µm. This value is negligible compared to the channel height; therefore, we assumed that the roughness did not affect the fluid flow. The PMMA layer with the fs laser-machined serpentine channel was coupled with a flat and smooth PMMA substrate with inlet and outlet holes drilled using the micromilling machine.

For both the devices, the next step in the fabrication process was the bonding of the PMMA layers. For the assembly of the microfluidic device in both experiments, a thermal- and solvent-assisted bonding method was implemented. In a protected environment we spin-coated hot isopropyl alcohol on the surface of the substrates, aligned the wet slices, and transferred the devices into the oven by holding them in position with clamps and creating an irreversible bonding. In order to build the multilayered chip, it was necessary to reiterate this process twice, i.e., for bonding the substrates with the microchannels, for placing the substrate with the holes on top of the channels. In the end, the channels with the interconnection hole were buried, while the inlets and outlets remained on the upper layer, thus resulting in a monolithic device assembled with no need for additional glue, luer, or gaskets. The two different studies, for mixing of solutions and capturing of CTC, required a diverse functionalization of the microchannels.

The 3D micromixer underwent O_2 plasma treatment to improve hydrophilicity. After that, the bonding quality and the existence of any leakage were examined by connecting the device to the Elvesys micropumping system through capillary tubes and by gradually increasing the pressure from 10 to 800 mbar. An optical microscope was used for the evaluation. The last step during this process was the in-flow functionalization.

2.4. Microchannel Passivation and Functionalization

Although PMMA is a good alternative for customizing the design of microfluidic modules, it suffers from high hydrophobicity. Thus, to attain an optimum functionality of our device, it was essential to mitigate the hydrophobicity of the PMMA surface. To this end, the process could be initiated by treating the assembled PMMA slices with O_2 plasma, instigating an improved surface wettability and hydrophilicity for an easier flow of water-based solutions. In the subsequent phase, in order to attenuate any sticking of the particles into the channel surface, which could impact the performance, it was necessary to incubate the chip for 2 h with blocking buffer (1 mg/mL BSA in PBS). The sample's injection during the functionalization process was done directly in-flow. This was mostly obtained thanks to the perfect fitting between the holes and the capillary tubes, allowing for a plug-and-play usage of the device. The stable connections did not require additional glue, gaskets, or clamps to avoid any leakage of the solution, apart from the capillary tubes and the channels.

In the case of the device for cancer cell capturing, it was necessary to flow a sequence of solutions into the microfluidic chip to functionalize it, starting from APTES (5%) in ethanol, in order to increase the hydrophilization and the amine functionality of the PMMA surface. This step was essential both for the easier flowing of the water-based solutions and for promoting the next step. Subsequently, after cleaning the surface with pure water, we injected glutaraldehyde (0.05%) to allow antibody immobilization. Specifically, the glutaraldehyde underwent an imine coupling reaction with the amine group of the antibody, resulting in immobilization of the anti-EpCAM antibody. The ultimate step of this process was the incubation of the device with a blocking buffer (BSA-Tween®20 in PBS), to prevent any cell absorption. Moreover, during this process, the device was connected with the micropumping system through a perfect fitting between the capillary tubes and the micromilled inlets.

2.5. Experimental Tests for Particles and Coloured Liquids

In order to gain better insight into the performance of the micromixer, the mixing behavior of the device was investigated. Thus, experiments for mixing and gradient generation were performed using colored fluids and particles. We also examined the case when the outlet transmuted into an inlet, to detect the mixing capabilities of the chip and the influence of different paths, lengths, and shapes of the microchannels. Our first attempt was to mix two different colored liquids. For this, we used the capillary tubes to simultaneously inject the liquids into the serpentine and reservoir channels. As a second checkpoint, we chose two different kinds of particles with diameters of 200 nm and 1 μm. Both experiments were performed at different flow rates.

For the injection of both samples, we connected the chip with the Elveflow microfluidic set up. The setup was equipped with an OB1 base module, two MkIII+ channels for pressure control, and two microfluidic sensors, with an analogous temporal flow control. The two inlets were connected with two different vials, containing either the colored liquids or the particles samples. For evaluating the mixing quality, the microchip was placed under a microscope, enabling real-time evaluation of the flow behaviors, as well as image acquisition.

2.6. Experimental Tests for Cells

In the CTC capturing experiment, we grew cells in an incubator at 37 °C with 5% CO_2, in suspension in RPMI 1640 complete growth medium and in adhesion in complete Dulbecco's modified Eagle's medium for Jurkat cells and OECM-1 cells, respectively. The growth medium was renewed every 2 days. A few minutes before the experiment started, cancer cells were suspended and harvested in 0.05% trypsin.

For sample injection into the device, we used the Elveflow micropumping system. We visualized the capturing with real-time image acquisition using an Axiozoom Zeiss V16 fluorescence microscope.

Finally, after detachment, the cells were washed and resuspended in DMEM medium. Jurkat cells were centrifuged, counted, and resuspended at the right dilution in order to inject them into the microfluidic chip. Flow rate was tuned at 7 μL/min.

Once the cells remained in contact with the microchannel walls, the microfluidic chip was gently washed with PBS, and adhered cells were stained with a subsequent injection of (i) anti-EpCAM antibody and (ii) secondary FITC-labeled antibody, in order to identify tumor cells blocked at the channels' surface.

3. Results and Discussion

3.1. Verification of the Numerical Model

The 3D fluidic micromixer was studied using CFD simulations, under different regimes, i.e., mixing of fluidics, and mixing and dilution of particles. The results of the interconnected microchannels for the flow velocity distribution and the pressure (Figure 2a,b) offered us the opportunity to gain better insight into the flow behavior. Specifically, Figure 2a depicts the velocity distribution before and after the mixing point, showing a maximal flow at the junction point and the minimal flow predominating at the inlets up to the point of convergence. The velocity at the side walls appeared to be infinitesimal. Concomitantly, the levels of pressure at the entire design are reported in the Figure 2b. As can be seen, the maxima and minima pressure levels were inversely proportional to the velocity. Subsequently, the flow rate was calculated using Equation (1).

$$Q = Au^{-}, \quad (1)$$

where Q is the flow rate, A indicates the cross-sectional area, and u is the average velocity.

To reveal the particle mixing behavior, a suspension of particles with sizes 1 μm and 200 nm was simulated and injected in both inlets. Moreover, in order to find the best combination of flow rates from the two inlets, we tested a range of values ranging from 1 μL/min to 5 μL/min. The recorded pressure was varied as a consequence of this variation. The mixing of the two different populations is illustrated in Figure 2c–f in the time range of 0–67 s.

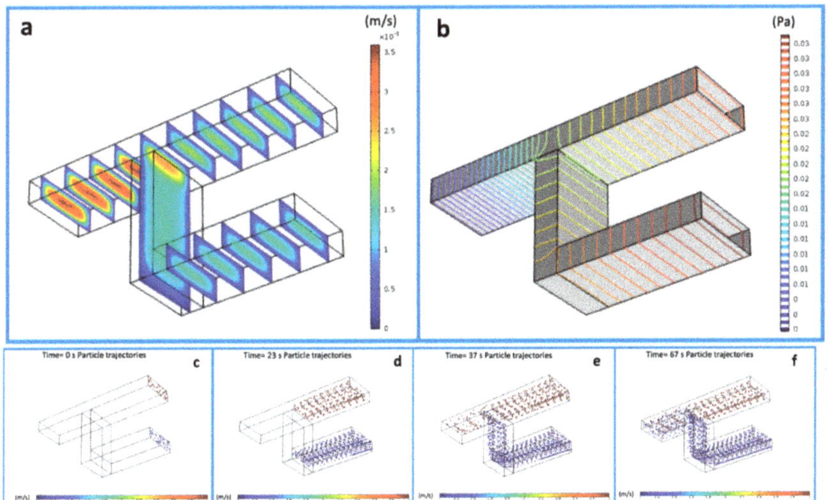

Figure 2. Simulated flow velocity (**a**) and pressure (**b**) in proximity of the interconnection point of the micromixing channel. Mixing of the two populations of nanoparticles at different time points, from the beginning of the experiment (**c**) at time 0 to the complete mixing obtained by (**d**): 23 seconds, (**e**): 37 seconds and accomplished after 67 seconds (**f**).

Design variables, such as the final shape of the microchannels containing a mixing chamber, were investigated. Specifically, we drew and simulated microchannels in rhomboidal, rectangular, and elliptical shapes in order to find the optimal one. Figure 3 shows the comparison of the fluidic velocity for the three different geometries: rectangle (3a), rhombus (3b), and oval (3c). A shared outcome among all cases was the high velocity in the center of the device and lower velocity at the edges. This condition was enhanced for the rectangular shape. Moreover, it was proven from experimental results that angular edges (as can be identified in the rhomboidal and rectangular shapes) are more prone to accumulate bubbles than round edges [55,56]. Thus, in our final device, we incorporated the elliptical shape for realizing the microfluidic reservoir.

3.2. Design and Fabrication of the LOC Devices

The findings of our simulation study in predicting the mixing process were verified by fabricating a micromixer with the selected oval and serpentine design. It was of crucial importance to select the optimal design parameters that had the greatest influence on the mixing quality. The first step for the realization of the device was to draw the CAD file and to transfer it into machine code for its fabrication through the micromilling machine. The proposed geometry for mixing experiments, as schematically illustrated in Section 2.3, contained a 7 cm long serpentine-shaped channel in the center of the device organized in six loops, with a reservoir-shaped channel in the bottom with a total length of around 5 cm. The top layer contained three holes (diameter: 1.8 mm) which were defined as inlets and outlets for the capillary tubes. Furthermore, the layer with the serpentine channel also featured a buried hole, which served as a junction point between the channels on the bottom layer and the top layer, thereby creating a 3D microfluidic pathway. The common portion after the junction point ran for 1.5 cm. The device was constructed/assembled from the three individual levels which were separately fabricated, using a mechanical micromilling machine with a 200 µm tool.

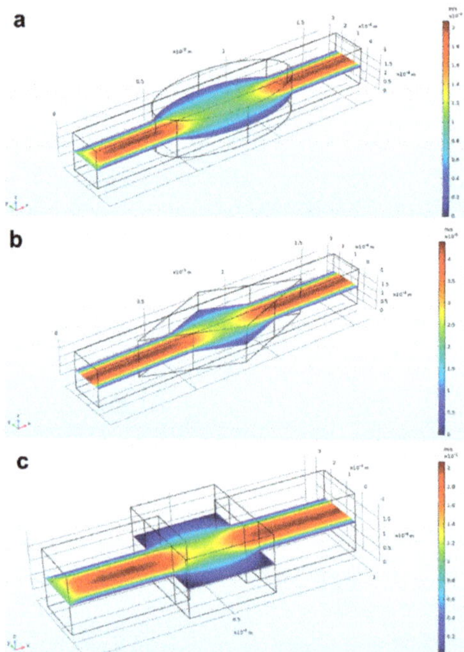

Figure 3. Simulated flow behavior into three different reservoirs: (**a**) rectangle; (**b**) rhombus; (**c**) oval. The lowest flow velocity occurred at the edges of (**a**,**b**).

Regarding the device used for cancer cell capturing, the layout was based on a serpentine microchannel with a square cross-section of 100 µm per side and a total length of 180 mm. The purpose of this design was to increase the active path and proportionally increase the possibility of capturing cells. As displayed in Figure 4a, the device consisted of two PMMA substrates. In this case, the upper substrate was micromilled in both faces. For the lower substrate, we exploited fs laser technology to fabricate the serpentine-shaped channel. To drill the inlet and outlet, we again used a mechanical micromilling machine, this time with a 400 µm tool, ensuring tight connections, since it fit perfectly with the capillary tubes and gave the opportunity for plug-and-play connections. The connection between the holes and the serpentine channel was achieved using two auxiliary channels with a diameter of 600 µm and length of 5 mm, fabricated on the same substrate. Lastly, the bottom PMMA flat layer allowed the sealing of the serpentine channel. No cracks, burrs, or recast layers stemmed from the microfabrication process, thus also providing a great transparency (Figure 4b). Furthermore, the roughness of the bottom channel (Ra = 2 µm) did not affect the fluidic transport of the cells since it was negligible compared to the channel's height.

The device was assembled and functionalized as described in Sections 2.3 and 2.4, and the possibility of real-time monitoring of flow into the microchannels, combined with the serpentine shape and a slow flow rate (2 µL/min), enabled us to attain a tool for exploiting a very high surface/volume ratio in terms of active binding sites for antibodies.

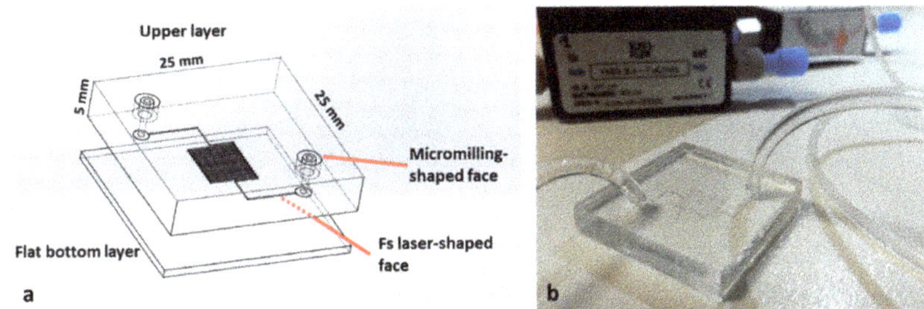

Figure 4. Features of the cell capture device (**a**). PMMA microfabrication allowed for complete transparency of the device. Inlets and outlets were designed to perfectly fit with capillary tubes (**b**).

3.3. Mixing and Gradient Generation Experiments

We characterized the sealing of the micromixer by visualizing the mixing behavior using two different colored liquids (Figure 5a). The progressive filling of the microchannels was supervised using a microscope. As can be visualized in Figure 5b, the paths of the 3D microchannels could be observed with a single microscope frame, allowing the contemporary monitoring of the multilevel structure. It was of crucial importance to make sure that both solutions arrived simultaneously at the mixing point; hence, we initially set the flow rate at 3 mL/min, but adjusted it later on, when needed. In particular, the complete transparency of the device allowed monitoring the channels while they were progressively filled and differently tuning the parameters due to the diverse shape, resistance, and velocity of the flow in each path. We noticed that the best combination for synchronized arrival at the junction was to set the flow rate to 2.08 μL/min and the pressure to 42.80 mbar for the serpentine channel, whereas these values were set to 1.65 μL/min and 63.08 mbar for the reservoir. The distribution of liquids in the channels and how they initially flowed independently (the pink solution in the upper serpentine channel and the blue one at the reservoir), before being mixed at the meeting point and assuming a violet color, clearly demonstrated that the buried hole (indicated by an arrow) interconnected the two fluids, which became indistinguishable after turning violet. Furthermore, the chip could be used as gradient generator tool using the inlet and outlet alternatively, by tuning the flow rates of the channels and establishing dominance of either the pink or the blue solution (Figure 5c–f).

We then assessed the utility of our apparatus using two test samples, where each one contained green fluorescent particles of 200 nm size with 9.1×10^5/mL concentration and red fluorescent particles of 1 μm size and 7.2×10^5/mL concentration, and we injected them into the reservoir and serpentine channel, respectively. The flow of each solution could run separately into the device as explained in the cartoon in Figure 6a. By using the fluorescence microscope, as can be seen in Figure 6b,c, the two different populations flowed separately (the greens were detected only in the bottom channel and the reds were detected only in the upper one). The final injecting parameters we applied in this experiment to simultaneously reach the common portion were a flow rate of 2.13 μL/min and 26.42 mbar pressure for the red particles and a flow rate of 1.46 μL/min and 39.52 mbar pressure for the green particles.

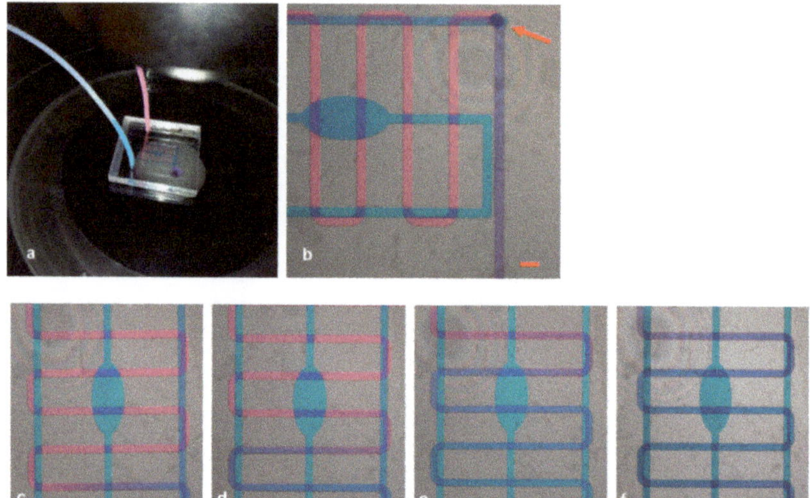

Figure 5. (**a**) Whole device connected to micropumps under the microscope; (**b**) frame acquired after complete mixing of pink and blue ink. (**c**–**f**) Modulation of flow rates, resulting in different mixing conditions of the inks. Scale bars: 500 μm.

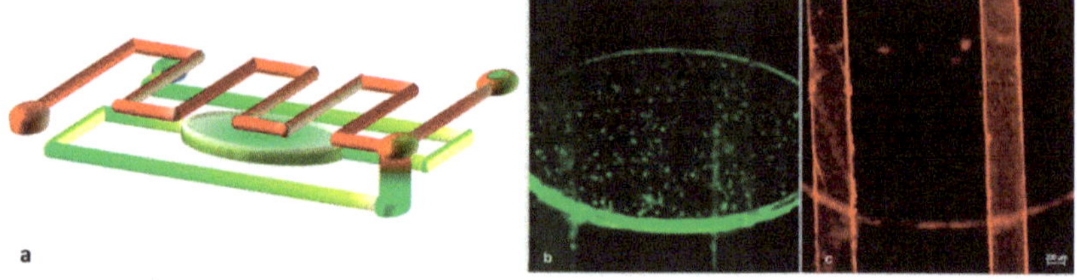

Figure 6. Solutions contained in the two channels flow separately into the device until they reach the interconnection point. (**a**) Scheme of the microchannel network filled with green and red nanoparticles. Green particles run in the bottom channel (**b**), while red ones run in the upper serpentine channel (**c**).

Successively, we performed the same experiment using a wide range of flow rates to test the performance of the chip. In all cases, to achieve a mutually proportional flow rate in both channels, every flow change in one channel was followed by a pressure adjustment in the other. Figure 7 presents the time-lapse images of particles captured soon before (Figure 7a–c) and after (Figure 7d–f) the mixing point. Here, the particles moved from their individual channels to the common one, thus verifying the expected mixing efficiency.

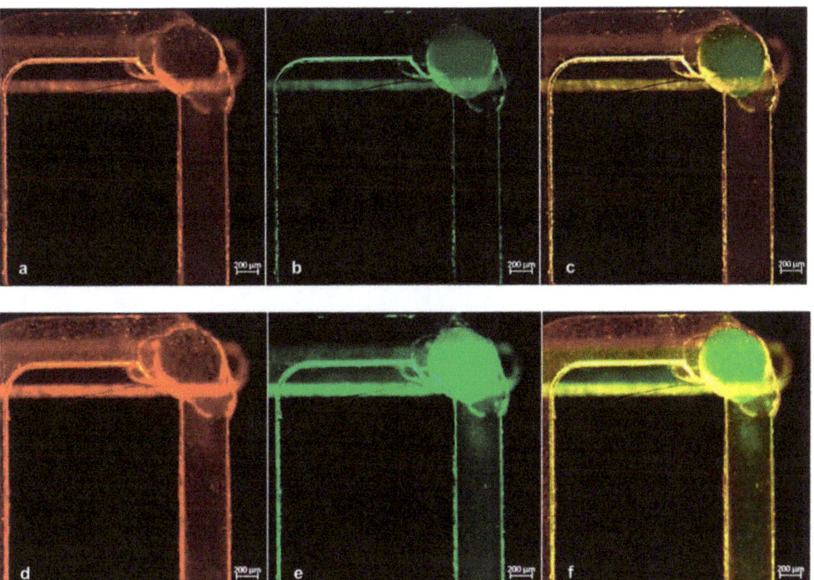

Figure 7. (**a**,**b**) Images of microchannels with green and red particles while running separately before reaching the interconnection point and (**d**,**e**) immediately after. (**c**,**f**) Merged images of green and red fluorescence acquisitions.

3.4. CTC Capture Experiments

An extensive test was also performed to evaluate the ability of the device to distinguish cancer cells from blood cells. With this aim, we used and immobilized anti-EpCAM antibodies able to recognize human EpCAM, which is a membrane biomarker typically present on the surface of tumoral epithelial cells. The immobilization procedure described in Section 3.3 resulted in the possibility of PMMA microchannels working as capture sites for oral cancer cells. Hence, as a proof of concept, we created a cell mixture composed of two different populations: Jurkat cells (blood-derived cells) and the OECM-1 human oral squamous carcinoma cell line (epithelial-like cells from human oral cancer). In this way, we were able to simulate the contents of a real complex sample. The two different samples were prepared separately and contained 1×10^6 cells/mL from the Jurkat line and 1×10^4 cells/mL from the OECM line. We injected the cell suspensions slowly through the serpentine channel with a flow rate of 7 µL/min. As Figure 8 displays, cells were recognized and blocked as long as they were expressing the EpCAM antigen on their membrane. Therefore, OECM-1 cells were captured on the inner walls of the channels, and most of them remained after the washing with PBS. To maximize the possibility of interaction of cells and wall channels, we used a very low flow rate (2 µL/min) to inject the cell suspension into the serpentine path. Moreover, PMMA device with its complete transparency provided sufficient proof of concept for the successful distinction of cancer cells from normal blood cells and their immobilization in a label-free manner. In the bright-field images of Figure 8, cells fixed to the microchannels walls are highlighted by a green spot.

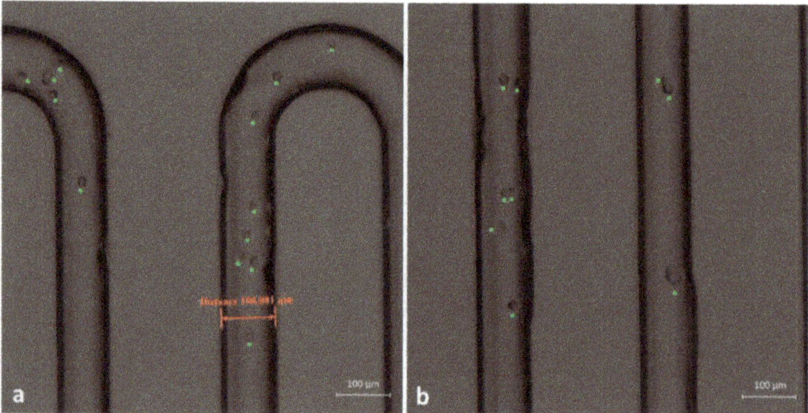

Figure 8. (**a**,**b**) Two frames in bright-field acquisition related to two different regions of the capture device. Attached cells are clearly recognizable and are highlighted by green crosses.

In order to demonstrate that captured cells were tumoral cells, we performed some additional labeling. Once the cells were blocked at the microchannels walls, we again injected anti-EpCAM antibody solution, after which we flowed a solution of secondary FITC-labeled antibody, able to bind the Fc portion of the primary antibody. In this way, we were able to selectively label the cell membrane of oral cancer cells. As visible in Figure 9, the membrane of fixed cells was labeled with fluorescent green antibody. We can, thus, conclude that the functionalized microchannels were able to selectively capture tumor cells.

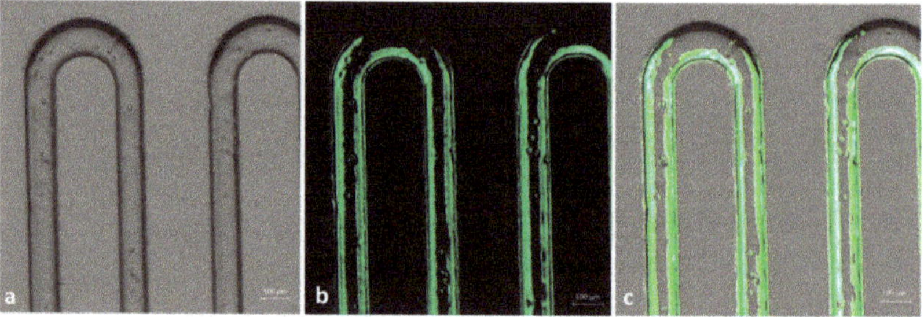

Figure 9. Identification of OECM fixed cells with anti-EpCAM antibody and secondary FITC-labeled antibody. (**a**) Bright-field acquisition of cells; (**b**) green fluorescent acquisition; (**c**) merged image of (**a**,**b**).

4. Conclusions

Plug-and-play devices for the rapid diagnosis of diseases are on the rise for the possibility to obtain quick results in a low-cost and easy-to-use manner. This paper, focusing on the development of a plastic disposable tool, describes the possibility of combining different functionalities, proposing a single chip able to stabilize, preserve, and prepare biological samples. The platform includes a module for sample preparation and mixing of solutions and a detection module, which was demonstrated to capture circulating tumor cells. The produced devices were fabricated via a highly customizable combination of fs laser and micromilling methods using low-cost plastic substrates, while allowing easy connections to the microfluidic system for in-flow functionalization and sample manipulation. The proposed functionalization is a proof of concept which can, in principle, be applied to the detection of other biological entities (exosomes, microvesicles, and so on), by simply modifying the antibody immobilized on the surface of PMMA microchannels. In this case, the validity of the assays was confirmed by using fluorescent probes, which in turn identified the micromixing of nanoparticles and the selective binding of tumor cells in a mixture of normal and cancer cells. The proposed devices may also be of great importance in the case of cancer cell investigations from other body fluids, e.g., saliva, which in turn may require preliminary steps for sample manipulation/dilution or reagent addition. These features, in the era of COVID-19, are very important; for example, a recent release from the American Food and Drug Administration (FDA) authorized the use of home-collected saliva to detect SARS Cov-2. In this way, patients are allowed to self-collect samples for analysis in order to improve accessibility to COVID-19 testing and decrease the risks of infection for medical personnel. Moreover, automatic and low-cost devices, in pandemic contexts, have the possibility to minimize interactions between patients and medical personnel, thus furtherly lowering the probability of infections without affecting access to large-scale screening programs for cancer (and other diseases).

Author Contributions: Planning and design, M.S.C. and F.F.; experimental activity, S.Z., A.V., C.G., M.S.C. and F.F.; experimental setup, A.A., F.F. and M.S.C.; implementation of the research, main conceptual ideas, and proof, E.P., M.S.C. and F.F.; analysis of the results, M.S.C., E.P., F.F., A.V. and A.A.; writing of the manuscript, S.Z., M.S.C., A.V. and C.G.; manuscript revision, S.Z., A.V., M.S.C., C.G., E.P., A.A. and F.F.; project supervision and funding responsibilities, M.S.C. and F.F. All authors have read and agreed to the published version of the manuscript.

Funding: This study was supported by the following funding programs: "SMILE (SAW-MIP Integrated Device for Oral Cancer Early Detection) project, part of the ATTRACT program funded by the European Union's Horizon 2020 Research and Innovation program (grant agreement: 777222)"; PRIN 2017 Project "Prostate cancer: disentangling the relationships with tumor microenvironment to better model and target tumor progression" (grant number: Prot. 20174PLLYN); Progetto PON ARS01_00906 "TITAN—Nanotecnologie per l'immunoterapia dei tumori", finanziato dal FESR nell'ambito del PON "Ricerca e Innovazione" 2014–2020—Azione II-OS 1.b).

Institutional Review Board Statement: Not applicable.

Informed Consent Statement: Not applicable.

Data Availability Statement: Not applicable.

Conflicts of Interest: The authors declare no conflict of interest.

References

1. Siegel, R.L.; Miller, K.D.; Jemal, A. Cancer statistics, 2020. *CA A Cancer J. Clin.* **2020**, *70*, 7–30. [CrossRef] [PubMed]
2. Vergaro, V.; Baldassarre, F.; de Santisa, F.; Ciccarella, G.; Giannelli, G.; Leporatti, S. TGF-beta inihibitor-loaded polyelectrolyte multilayers capsules for sustained targeting of hepatocarcinoma cells. *Curr. Pharm. Des.* **2012**, *18*, 4155–4164. [CrossRef]
3. Gupta, B.; Johnson, N.W.; Kumar, N. Global Epidemiology of Head and Neck Cancers: A Continuing Challenge. *Oncology* **2016**, *91*, 13–23. [CrossRef]
4. Hashibe, M.; Brennan, P.; Benhamou, S.; Castellsague, X.; Chu, C.; Curado, M.P.; Dal Maso, L.; Dauct, A.W.; Fabianova, E.; Wunsch, V.; et al. Alcohol drinking in never users of tobacco, cigarette smoking in never drinkers, and the risk of head and neck cancer: Pooled analysis in the international head and neck cancer epidemiology consortium. *J. Natl. Cancer Inst.* **2007**, *99*, 777–789. [CrossRef]
5. Perot, P.; Falguieres, M.; Arowas, L.; Laude, H.; Foy, J.P.; Goudot, P.; Corre-Catelin, N.; Ungeheuer, M.N.; Caro, V.; Heard, I.; et al. Investigation of viral etiology in potentially malignant disorders and oral squamous cell carcinomas in non-smoking, non-drinking patients. *PLoS ONE* **2020**, *15*. [CrossRef]
6. Petersen, P.E.; Bourgeois, D.; Ogawa, H.; Estupinan-Day, S.; Ndiaye, C. The global burden of oral diseases and risks to oral health. *Bull. World Health Organ.* **2005**, *83*, 661–669. [PubMed]
7. Li, L.; Zhang, Z.T. Genetic Association between NFKBIA and NFKB1 Gene Polymorphisms and the Susceptibility to Head and Neck Cancer: A Meta-Analysis. *Dis. Markers* **2019**, *2019*. [CrossRef]
8. Tsompana, M.; Gluck, C.; Sethi, I.; Joshi, I.; Bard, J.; Nowak, N.J.; Sinha, S.; Buck, M.J. Reactivation of super-enhancers by KLF4 in human Head and Neck Squamous Cell Carcinoma. *Oncogene* **2020**, *39*, 262–277. [CrossRef]
9. Ogino, S.; Nowak, J.A.; Hamada, T.; Milner, D.A., Jr.; Nishihara, R. Insights into Pathogenic Interactions among Environment, Host, and Tumor at the Crossroads of Molecular Pathology and Epidemiology. *Annu. Rev. Pathol.* **2019**, *14*, 83–103. [CrossRef] [PubMed]

10. Mima, K.; Kosumi, K.; Baba, Y.; Hamada, T.; Baba, H.; Ogino, S. The microbiome, genetics, and gastrointestinal neoplasms: The evolving field of molecular pathological epidemiology to analyze the tumor-immune-microbiome interaction. *Hum. Genet.* **2021**, *140*, 725–746. [CrossRef] [PubMed]
11. Shiga, K.; Tateda, M.; Saijo, S.; Hori, T.; Sato, I.; Tateno, H.; Matsuura, K.; Takasaka, T.; Miyagi, T. Presence of Streptococcus infection in extra-oropharyngeal head and neck squamous cell carcinoma and its implication in carcinogenesis. *Oncol. Rep.* **2001**, *8*, 245–248. [CrossRef]
12. Zhang, L.; Liu, Y.; Zheng, H.J.; Zhang, C.P. The Oral Microbiota May Have Influence on Oral Cancer. *Front. Cell. Infect. Microbiol.* **2020**, *9*, 11. [CrossRef] [PubMed]
13. Sinevici, N.; Mittermayr, S.; Davey, G.P.; Bones, J.; O'Sullivan, J. Salivary N-glycosylation as a biomarker of oral cancer: A pilot study. *Glycobiology* **2019**, *29*, 726–734. [CrossRef] [PubMed]
14. Zhang, C.Z.; Cheng, X.Q.; Li, J.Y.; Zhang, P.; Yi, P.; Xu, X.; Zhou, X.D. Saliva in the diagnosis of diseases. *Int. J. Oral Sci.* **2016**, *8*, 133–137. [CrossRef] [PubMed]
15. Vesty, A.; Gear, K.; Biswas, K.; Radcliff, F.J.; Taylor, M.W.; Douglas, R.G. Microbial and inflammatory-based salivary biomarkers of head and neck squamous cell carcinoma. *Clin. Exp. Dent. Res.* **2018**, *4*, 255–262. [CrossRef]
16. Diesch, T.; Filippi, C.; Fritschi, N.; Filippi, A.; Ritz, N. Cytokines in saliva as biomarkers of oral and systemic oncological or infectious diseases: A systematic review. *Cytokine* **2021**, *143*. [CrossRef]
17. Song, X.; Yang, X.; Narayanan, R.; Shankar, V.; Ethiraj, S.; Wang, X.; Duan, N.; Ni, Y.-H.; Hu, Q.; Zare, R.N. Oral squamous cell carcinoma diagnosed from saliva metabolic profiling. *Proc. Natl. Acad. Sci. USA* **2020**, *117*, 16167–16173. [CrossRef]
18. Day, A.T.; Fakhry, C.; Tiro, J.A.; Dahlstrom, K.R.; Sturgis, E.M. Considerations in Human Papillomavirus-Associated Oropharyngeal Cancer Screening A Review. *JAMA Otolaryngol-Head Neck Surg.* **2020**, *146*, 656–664. [CrossRef]
19. Tenore, G.; Nuvoli, A.; Mohsen, A.; Cassoni, A.; Battisti, A.; Terenzi, V.; Della Monaca, M.; Raponi, I.; Brauner, E.; De Felice, F.; et al. Tobacco, Alcohol and Family History of Cancer as Risk Factors of Oral Squamous Cell Carcinoma: Case-Control Retrospective Study. *Appl. Sci. Basel* **2020**, *10*, 11. [CrossRef]
20. Zoupanou, S.; Chiriacò, M.S.; Tarantini, I.; Ferrara, F. Innovative 3D Microfluidic Tools for On-Chip Fluids and Particles Manipulation: From Design to Experimental Validation. *Micromachines* **2021**, *12*, 104. [CrossRef] [PubMed]
21. Volpe, A.; Krishnan, U.; Chiriacò, M.S.; Primiceri, E.; Ancona, A.; Ferrara, F. A smart procedure for the femtosecond laser-based fabrication of a polymeric lab-on-a-chip for capturing tumor cell. *Engineering* **2020**, *12*. [CrossRef]
22. Trotta, G.; Volpe, A.; Ancona, A.; Fassi, I. Flexible micro manufacturing platform for the fabrication of PMMA microfluidic devices. *J. Manuf. Process.* **2018**, *35*, 107–117. [CrossRef]
23. Volpe, A.; Paie, P.; Ancona, A.; Osellame, R. Polymeric fully inertial lab-on-a-chip with enhanced-throughput sorting capabilities. *Microfluid. Nanofluidics* **2019**, *23*. [CrossRef]
24. Niculescu, A.G.; Chircov, C.; Birca, A.C.; Grumezescu, A.M. Fabrication and Applications of Microfluidic Devices: A Review. *Int. J. Mol. Sci.* **2021**, *22*, 2011. [CrossRef]
25. Abgrall, P.; Gue, A.M. Lab-on-chip technologies: Making a microfluidic network and coupling it into a complete microsystem-A review. *J. Micromech. Microeng.* **2007**, *17*, R15–R49. [CrossRef]
26. Abdelgawad, M.; Watson, M.W.L.; Young, E.W.K.; Mudrik, J.M.; Ungrin, M.D.; Wheeler, A.R. Soft lithography: Masters on demand. *Lab Chip* **2008**, *8*, 1379–1385. [CrossRef] [PubMed]
27. Wilson, M.E.; Kota, N.; Kim, Y.; Wang, Y.; Stolz, D.B.; LeDuc, P.R.; Ozdoganlar, O.B. Fabrication of circular microfluidic channels by combining mechanical micromilling and soft lithography. *Lab Chip* **2011**, *11*, 1550–1555. [CrossRef] [PubMed]
28. Chiriaco, M.S.; Bianco, M.; Amato, F.; Primiceri, E.; Ferrara, F.; Arima, V.; Maruccio, G. Fabrication of interconnected multilevel channels in a monolithic SU-8 structure using a LOR sacrificial layer. *Microelectron. Eng.* **2016**, *164*, 30–35. [CrossRef]
29. Hardy, B.S.; Uechi, K.; Zhen, J.; Pirouz Kavehpour, H. The deformation of flexible PDMS microchannels under a pressure driven flow. *Lab Chip* **2009**, *9*, 935–938. [CrossRef] [PubMed]
30. Bhattacharjee, N.; Urrios, A.; Kanga, S.; Folch, A. The upcoming 3D-printing revolution in microfluidics. *Lab Chip* **2016**, *16*, 1720–1742. [CrossRef]
31. Yazdi, A.A.; Popma, A.; Wong, W.; Nguyen, T.; Pan, Y.Y.; Xu, J. 3D printing: An emerging tool for novel microfluidics and lab-on-a-chip applications. *Microfluid. Nanofluidics* **2016**, *20*. [CrossRef]
32. Available online: https://ttconsultants.com/3d-printing-limitations/ (accessed on 23 July 2021).
33. Farson, D.F.; Choi, H.W.; Zimmerman, B.; Steach, J.K.; Chalmers, J.J.; Olesik, S.V.; Lee, L.J. Femtosecond laser micromachining of dielectric materials for biomedical applications. *J. Micromech. Microeng.* **2008**, *18*, 035020. [CrossRef]
34. Volpe, A.; Gaudiuso, C.; Ancona, A. Sorting of Particles Using Inertial Focusing and Laminar Vortex Technology: A Review. *Micromachines* **2019**, *10*, 594. [CrossRef] [PubMed]
35. Kononenko, T.V.; Konov, V.I.; Garnov, S.V.; Danielius, R.; Piskarskas, A.; Tamosauskas, G.; Dausinger, F. Comparative study of the ablation of materials by femtosecond and pico- or nanosecond laser pulses. *Quantum Electron.* **1999**, *29*, 724–728. [CrossRef]
36. Volpe, A.; Covella, S.; Gaudiuso, C.; Ancona, A. Improving the Laser Texture Strategy to Get Superhydrophobic Aluminum Alloy Surfaces. *Coatings* **2021**, *11*, 369. [CrossRef]
37. Ashkenasi, D.; Müller, G.; Rosenfeld, A.; Stoian, R.; Hertel, I.V.; Bulgakova, N.M.; Campbell, E.E.B. Fundamentals and advantages of ultrafast micro-structuring of transparent materials. *Appl. Phys. A* **2003**, *77*, 223–228. [CrossRef]

38. Faustino, V.; Catarino, S.O.; Lima, R.; Minas, G. Biomedical microfluidic devices by using low-cost fabrication techniques: A review. *J. Biomech.* **2016**, *49*, 2280–2292. [CrossRef]
39. Guckenberger, D.J.; de Groot, T.E.; Wan, A.M.D.; Beebe, D.J.; Young, E.W.K. Micromilling: A method for ultra-rapid prototyping of plastic microfluidic devices. *Lab Chip* **2015**, *15*, 2364–2378. [CrossRef] [PubMed]
40. Song, I.-H.; Park, T. Connector-Free World-to-Chip Interconnection for Microfluidic Devices. *Micromachines* **2019**, *10*, 166. [CrossRef]
41. Yuen, P.K. A reconfigurable stick-n-play modular microfluidic system using magnetic interconnects. *Lab Chip* **2016**, *16*, 3700–3707. [CrossRef]
42. Berthier, E.; Young, E.W.K.; Beebe, D. Engineers are from PDMS-land, Biologists are from Polystyrenia. *Lab Chip* **2012**, *12*, 1224–1237. [CrossRef]
43. Rotem, A.; Abate, A.R.; Utada, A.S.; Van Steijn, V.; Weitz, D.A. Drop formation in non-planar microfluidic devices. *Lab Chip* **2012**, *12*, 4263–4268. [CrossRef]
44. Nguyen, T.; Chidambara, V.A.; Andreasen, S.Z.; Golabi, M.; Huynh, V.N.; Linh, Q.T.; Bang, D.D.; Wolff, A. Point-of-care devices for pathogen detections: The three most important factors to realise towards commercialization. *TrAC Trends Anal. Chem.* **2020**, *131*. [CrossRef]
45. Temiz, Y.; Lovchik, R.D.; Kaigala, G.V.; Delamarche, E. Lab-on-a-chip devices: How to close and plug the lab? *Microelectron. Eng.* **2015**, *132*, 156–175. [CrossRef]
46. Vazquez, R.M.; Trotta, G.; Volpe, A.; Bernava, G.; Basile, V.; Paturzo, M.; Ferraro, P.; Ancona, A.; Fassi, I.; Osellame, R. Rapid Prototyping of Plastic Lab-on-a-Chip by Femtosecond Laser Micromachining and Removable Insert Microinjection Molding. *Micromachines* **2017**, *8*, 328. [CrossRef] [PubMed]
47. Chaudhuri, P.K.; Ebrahimi Warkiani, M.; Jing, T.; Kenry; Lim, C.T. Microfluidics for research and applications in oncology. *Analyst* **2016**, *141*, 504–524. [CrossRef] [PubMed]
48. Song, W.; Li, X.; Zhao, Y.; Liu, C.; Xu, J.; Wang, H.; Zhang, T. Functional, UV-curable coating for the capture of circulating tumor cells. *Biomater. Sci.* **2019**, *7*, 2383–2393. [CrossRef] [PubMed]
49. Sathish, S.; Ishizu, N.; Shen, A.Q. Air Plasma-Enhanced Covalent Functionalization of Poly(methyl methacrylate): High-Throughput Protein Immobilization for Miniaturized Bioassays. *Acs Appl. Mater. Interfaces* **2019**, *11*, 46350–46360. [CrossRef] [PubMed]
50. Wang, X.; Deng, L.H.; Gjertsen, B.T. A microfluidic device for differential capture of heterogeneous rare tumor cells with epithelial and mesenchymal phenotypes. *Anal. Chim. Acta* **2020**, *1129*, 1–11. [CrossRef]
51. Wu, X.; Bai, Z.; Wang, L.; Cui, G.; Yang, M.; Yang, Q.; Ma, B.; Song, Q.; Tian, D.; Ceyssens, F.; et al. Magnetic Cell Centrifuge Platform Performance Study with Different Microsieve Pore Geometries. *Sensors* **2020**, *20*, 48. [CrossRef]
52. Lim, L.S.; Hu, M.; Huang, M.C.; Cheong, W.C.; Gan, A.T.L.; Looi, X.L.; Leong, S.M.; Koay, E.S.-C.; Li, M.-H. Microsieve lab-chip device for rapid enumeration and fluorescence in situ hybridization of circulating tumor cells. *Lab Chip* **2012**, *12*, 4388–4396. [CrossRef]
53. To, T.D.; Truong, A.T.T.; Nguyen, A.T.; Doan, T.C.D.; Dang, C.M. Filtration of circulating tumour cells MCF-7 in whole blood using non-modified and modified silicon nitride microsieves. *Int. J. Nanotechnol.* **2018**, *15*, 39–52. [CrossRef]
54. Volpe, A.; Paie, P.; Ancona, A.; Osellame, R.; Lugara, P.M.; Pascazio, G. A computational approach to the characterization of a microfluidic device for continuous size-based inertial sorting. *J. Phys. D Appl. Phys.* **2017**, *50*. [CrossRef]
55. Pereiro, I.; Fomitcheva Khartchenko, A.; Petrini, L.; Kaigala, G.V. Nip the bubble in the bud: A guide to avoid gas nucleation in microfluidics. *Lab Chip* **2019**, *19*, 2296–2314. [CrossRef] [PubMed]
56. Blonski, S.; Zaremba, D.; Jachimek, M.; Jakiela, S.; Wacławczyk, T.; Korczyk, P.M. Impact of inertia and channel angles on flow distribution in microfluidic junctions. *Microfluid. Nanofluidics* **2020**, *24*, 14. [CrossRef]

 *micromachines*

Article

A New Predictive Technology for Perinatal Stem Cell Isolation Suited for Cell Therapy Approaches

Silvia Zia [1], Giulia Martini [2], Valeria Pizzuti [2,3], Alessia Maggio [2], Giuliana Simonazzi [4], Pierluigi Reschiglian [1,5], Laura Bonsi [2], Francesco Alviano [2], Barbara Roda [1,5] and Andrea Zattoni [1,5,*]

1 Stem Sel srl., 40127 Bologna, Italy; silvia.zia@stemsel.it (S.Z.); pierluigi.reschiglian@unibo.it (P.R.); barbara.roda@unibo.it (B.R.)
2 Unit of Histology, Department of Experimental, Diagnostic and Specialty Medicine, Embryology and Applied Biology, University of Bologna, 40126 Bologna, Italy; giulia.martini18@studio.unibo.it (G.M.); valeria.pizzuti3@unibo.it (V.P.); Alessia.maggio4@studio.unibo.it (A.M.); laura.bonsi@unibo.it (L.B.); francesco.alviano@unibo.it (F.A.)
3 Unit of Nephrology, Department of Experimental, Diagnostic and Specialty Medicine, Dialysis and Renal Transplant, St. Orsola-Malpighi University Hospital, 40138 Bologna, Italy
4 Obstetric Unit, Department of Medical and Surgical Sciences, Policlinico St. Orsola-Malpighi, University of Bologna, 40126 Bologna, Italy; giuliana.simonazzi@unibo.it
5 Department of Chemistry G. Ciamician, University of Bologna, 40126 Bologna, Italy
* Correspondence: andrea.zattoni@unibo.it

Abstract: The use of stem cells for regenerative applications and immunomodulatory effect is increasing. Amniotic epithelial cells (AECs) possess embryonic-like proliferation ability and multipotent differentiation potential. Despite the simple isolation procedure, inter-individual variability and different isolation steps can cause differences in isolation yield and cell proliferation ability, compromising reproducibility observations among centers and further applications. We investigated the use of a new technology as a diagnostic tool for quality control on stem cell isolation. The instrument label-free separates cells based on their physical characteristics and, thanks to a micro-camera, generates a live fractogram, the fingerprint of the sample. Eight amniotic membranes were processed by trypsin enzymatic treatment and immediately analysed. Two types of profile were generated: a monomodal and a bimodal curve. The first one represented the unsuccessful isolation with all recovered cell not attaching to the plate; while for the second type, the isolation process was successful, but we discovered that only cells in the second peak were alive and resulted adherent. We optimized a Quality Control (QC) method to define the success of AEC isolation using the fractogram generated. This predictive outcome is an interesting tool for laboratories and cell banks that isolate and cryopreserve fetal annex stem cells for research and future clinical applications.

Keywords: fetal stem cells; amniotic epithelial cells; isolation protocol; quality control; label-free sorting; diagnostic tool

Citation: Zia, S.; Martini, G.; Pizzuti, V.; Maggio, A.; Simonazzi, G.; Reschiglian, P.; Bonsi, L.; Alviano, F.; Roda, B.; Zattoni, A. A New Predictive Technology for Perinatal Stem Cell Isolation Suited for Cell Therapy Approaches. *Micromachines* **2021**, *12*, 782. https://doi.org/10.3390/mi12070782

Academic Editor: Nam-Trung Nguyen

Received: 24 May 2021
Accepted: 28 June 2021
Published: 30 June 2021

1. Introduction

Advanced therapy medicinal products (ATMP) are medicines based on genes, cells and tissues to treat human diseases. The somatic-cell therapy consists of cells infusion that replaces tissue functions, cures and prevents diseases. In the last decade, cell therapy approaches, and in particular stem cells (SCs) treatments, are increasing. Adult SCs are widely used to treat malignant diseases like leukemia by hematopoietic stem cells transplantation, and Graft Versus Host Disease (GVHD) by bone marrow mesenchymal stem cells (BM-MSCs) for their immunomodulatory capacity [1–5]. Among stem cell types, perinatal SCs have gained attention because they possess wide differentiation potential and tolerogenic ability [6]. Placenta is a rich source of stem cells: mesenchymal and epithelial cells with staminal characteristics can be derived and it was proven their therapeutic potential in various disease models [7]. Amniotic epithelial cells (AECs) derive from the innermost layer of the amniotic membrane, the one in direct contact with the amniotic fluid. They possess the ability to differentiate toward all three germ layers, they are not tumorigenic and they have immunosuppressive features. AECs are cuboidal epithelial cells firmly adherent to a thick basement membrane [8], expressing high levels of epithelial adhesion molecules, such as EpCAM (CD326) and integrin subunits (CD29 and CD49f), while lacking typical stromal markers expression [9]. AECs also express several pluripotency markers including octamer-binding protein 4 (OCT-4), SRY-related HMG-box gene 2 (SOX-2), and Nanog and showed multilineage differentiation capacity [10]. Perinatal cell populations, including AECs, are also characterized by physiological immunomodulatory and anti-inflammatory properties due to their embryonic origin [11]. Pre-clinical studies have shown AECs' therapeutic effect in neurological disorders, lung injury, liver injury, diabetes, acute kidney failure, cardiovascular diseases and wound healing among many [12–22]. Moreover, AECs have been proved safe and non-tumorigenic upon transplantation, with no expression of telomerase and limited growing potential in culture [10].

Despite their efficacy and safety, AECs expressed differences in stemness characteristics, differentiation potential and immunomodulatory activities depending on the heterogeneity of primary derived cells which lead to a variable effect based on population composition [23,24]. Thanks to their early origin and easy access, AECs could be isolated, cryopreserved in specialized cell banks for future autologous therapy approaches [25]. Despite AECs isolation method is easy and does not require expensive material, the yield of AECs can be quite variable and their characteristics appear to be dependent on the genotype, age of the donor, region of cell isolation on placenta [24], cross-contamination of amniotic epithelial and mesenchymal stromal cell, isolation protocol, epithelial-to-mesenchymal transition of AECs [26] and measuring methods that are used for characterization [27]. Few studies have focused attention on membrane microscopic observation before enzymatic treatment [28] but it is operator-dependent. Several protocols have been proposed for isolation of hAECs with a wide range of yielded cells, viability and purity [29,30].

Reproducible and accurate systems are needed to standardize the isolation protocol of primary SCs. Identification and selection procedures able to isolate stem cells are essential to most cell therapy models. Multiple methods have been developed including mechanical sorting, surface receptors or biological markers of stemness. All these processes involve the knowledge of specific marker expression, which is not always known, and cell manipulation that must be scalable and amenable to GMP procedures. These requirements may be no trivial. Therefore, development of new technologies and relevant application methods are welcomed.

Microfluidic systems are widely used to test quality of pharmaceutical compounds but the working dimensional range belongs to micro-nanoparticles [31]. In the last decade, field-flow fractionation (FFF) has proven its capacity to analyze, discriminate and separate a wide size range of cells mixture based on their physical characteristics with high resolution and throughput [32,33]. In order to work with cells that have the ability to adhere to plastic, a novel method has been developed, the Non-Equilibrium Earth Gravity Assisted Dynamic fractionation (NEEGA-DF) [34]. Adhesion and contact of cells with the fractionation device are totally avoided by in-flow injection, by the absence of stop-flow cell sedimentation, and by using elution flow rate values able to generate hydrodynamic forces that are intense enough to lift and keep cells away from the channel wall. Cells having different physical characteristics acquire different velocity inside the capillary channel and elute at different time, so it is possible to collect the subpopulations composing the biological sample. It was proven that mesenchymal and epithelial cells from different organs origin, showed specific profile outcome of the separation process meaning that this method is suitable to underlined intra-differences in the cell population that other techniques do not do [34]. We developed an automated instrument that implements the NEEGA-DF method (Celector®, Stem Sel srl, Bologna, Italy), using a micro-camera for cell detection and a specifically designed software for image acquisition, post-processing and data analysis. The output of the instrument is a multiparametric fractogram representing number, size and shape of the eluted cells as a function of fractionation time and it is the fingerprint of the cell sample.

In this study, we used the instrument Celector® to perform the quality check of freshly isolated AECs, to compare possible differences in cells' yield and composition of amniotic membrane treated with two concentrations of trypsin, 0.1% and 0.25%. The live fractogram was used as predictive data to define successful isolation procedure and additionally, post-processing image data were compared to biological data of cell recovery, cell vitality and adhesion ability.

2. Materials and Methods

2.1. Instrumentation

2.1.1. Fractionation Principle and Procedure

The separation is obtained in a capillary device (channel) with rectangular cross section where cells suspensions are eluted through a laminar flow of physiological buffer. When a cell suspension is injected at a flow rate of 1mL min into the system, cells are transported by the flow and reach a specific position across the channel thickness during transportation due to the combined action of gravity, acting perpendicularly to the flow, and opposing lift forces that depend on the morphological features of the sample. Cells at a specific position in the channel acquire well-defined velocities and are therefore eluted at specific times [35]. The in-flow injection, the absence of a stop-flow, cell sedimentation step, and the use of elution flow rate values able to generate hydrodynamic forces that are intense enough to lift and keep cells away from the channel wall, make cells avoid any contact with the device with a consequent maintenance of native properties and high sample recovery.

The fractionation procedure involves at first the decontamination of the fractionation system by flushing with cleaning solution at 1 mL min flow rate. Next, the system was washed copiously with sterile, demineralized water at the same flow rate. Although the NEEGA-DF method is optimized to prevent contact between cells and fractionation device, to block non-specific interaction sites on the plastic walls, the fractionation system was flushed at 0.5 mL min with a sterile coating solution. Finally, it was filled with a sterile mobile phase. All solutions were provided by Stem Sel srl. The instrument is placed under a laminar flow hood to maintain the sterility of the collected cells.

2.1.2. Optical Analysis

Eluted cells were monitored using a micro-camera detector (MER-U3 camera, DAHENG IMAGING, Beijing, China) that is placed at the outlet of the fractionation channel. The imaging software (Celector Optics, Stem Sel srl, Italy) generates real-time fractogram representing the percentage of frame area covered by the cells versus recorded time. The imaging data are then post processed to obtain number and geometrical features of eluted cells as a function of time. In this work, we focused on the area/diameter and circularity of cells to obtain information on population heterogeneity and composition of possible sub-populations. These geometrical features were then visualized as scatter plot and curves using dedicated data processing (Stem Sel Analyzer), to obtain the average of all parameters in a selected time interval (cell fraction).

2.2. Cell Analysis and Collection

For every sample, cells were first analyzed to obtain a patient-specific fractogram and identify the fractions to collect. Consecutive analyses were run to increase the number of collected cells per fraction. The fractionated cells were centrifuged, and cell recovery was calculated by erythrosine dye (Sigma, St. Louis, MO, USA) to count alive and dead cells for both groups, 0.1 and 0.25% trypsin.

2.3. Isolation of Human Amniotic Epithelial Cells (AECs)

The study was approved by the Local Research Ethics Committee (EM894-2020_68/2017/U/Tess/AOUBo) and written informed consent was obtained from each healthy donor before specimen collection.

Amnion membranes were retrieved from term pregnancies (37–40 weeks of gestation) delivered by Caesarean section. hAECs were isolated using a modified procedure of the protocol previously reported [36], the amnion layer was mechanically peeled off the chorion layer and immediately washed in Phosphate Buffer Saline (PBS, Gibco, ThermoFisher Scientific, Waltham, MA, USA) without calcium and magnesium (HBSS, PAA Laboratories GmbH, Pasching, Austria) until blood clots were completely removed. For each sample, the amnion was minced into 4 to 6 medium size pieces (25 cm^2 approximately) and divided into two groups: one group was treated with 0.1% of trypsin-EDTA and the other group with a 0.25% trypsin-EDTA for all digestion steps (Gibco). Before enzymatic treatment, membrane was washed in PBS and 0.5 μM EDTA for 10 min then firstly, pieces were incubated for 10 min at 37 °C to exclude debris and then incubated for a second and third 40 min enzymatic digestion using the two concentrations to release the amniotic epithelial cells (AECs). Single cell suspension was washed with PBS and tested for viability with erythrosine dye (Biochrom AG, Berlin, Germany) and the number of viable and dead cells were counted. Cellular pellets were resuspended in the growth medium consisted of DMEM high glucose supplemented with 10% FBS, EGF (10 ng/mL; Sigma-Aldrich, St. Louis, MO, USA) and 1% penicillin and streptomycin (all solutions from Gibco) and plated at a density of 100,000 cells/cm^2 in a T25 flask for expansion. The leftover cells, at least 1.2×10^6 cells, were analysed using Celector® for quality control of the isolation process and cell sorting.

For the study AECs from the second digestion were diluted to a final concentration of 3×10^6 cells per mL and 100 μL were injected. Cells were automatically re-dispersed 3 times to homogenize the suspension and eluted at a flow rate of 1 mL min.

2.4. Downstream Analysis

Collected cells for each fraction and from the control group were seeded at a cell density of 100,000 cells/cm^2 to observe cell adhesion ability and morphology. In order to define the success of the isolation protocol, cells must adhere to plastic surface and show proliferative ability one week after isolation occur. If cells did not adhere or showed no ability in proliferation, isolation was defined as unsuccessful.

10,000 freshly sorted cells of each group, 0.1 and 0.25% trypsin, were seeded on a glass coverslip. After 4 days, cells were fixed in 10% formalin and stained for nuclear DAPI (Prolonged antifade, Molecular Probes). Images were taken using a fluorescent microscope and analyzed using the NII plugin to determine the number of normal, senescent and mitotic nuclei [37].

2.5. Statistical Analysis

Statistical analysis was performed using Graph Pad Prism v 8, running the *t*-test and mean and standard deviation were graphed.

3. Results

3.1. Predictivity

Amniotic epithelial cells (AECs) were isolated using different concentration of trypsin, 0.1 and 0.25%. Freshly isolated cells from the second digestion were immediately analysed by Celector® to profile populations. Two types of profiles were generated: a profile having two distinct peaks (type 1) (Figure 1A) and a profile with all cells eluted in the first minute of the analysis (type 2) (Figure 1C). Type 1 profile represents alive and proliferating cells while type 2 showed unsuccessful protocol, with no cells attaching to the plate and not able to proliferate. For both type of profile, unretained cells and debris eluted in the first minute of the analysis, then first subpopulation eluted between the 3rd and the 7th minute of analysis (Fraction 1, F1), and when present the second sub-population eluted between the 7th and the 14th minute (Fraction 2, F2). No difference was observed between the use of different trypsin concentration; both concentrations showed the same predictivity about the isolation protocol. Cell aggregates were observed in F1 for both trypsin treatments, while single cells eluted in F2 (Figure 1B). Sample belonging to type 2-profile, the one resulting unsuccessful in the cell expansion, presented more and bigger cell aggregates in F1 compared to type 1 samples (Figure 1D) and very few cells eluted in F2.

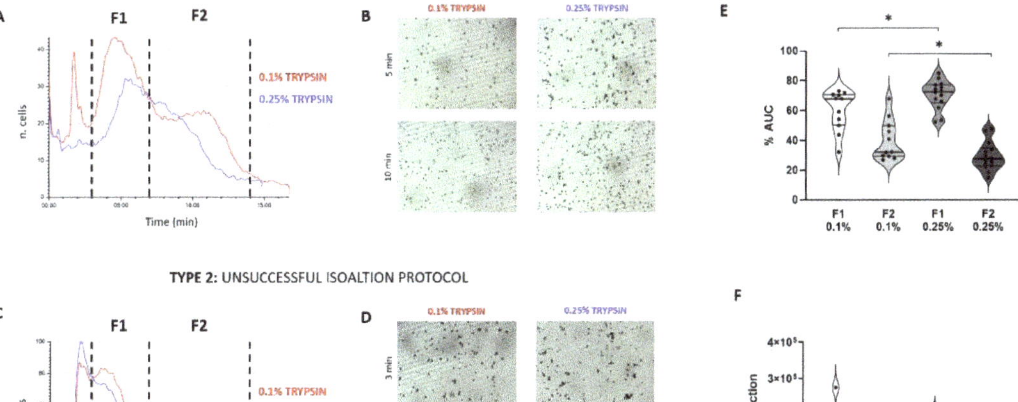

Figure 1. Representative images of a successful isolation protocol of AECs (Type 1) and an unsuccessful (Type 2) protocol. Profile represents the number of cells versus time of analysis (**A**,**C**) for both trypsin treatment 0.1 and 0.25%. The time interval of cells collection is shown as a dotted line that divides the two subpopulations F1 and F2 (F1 from the 3rd to the 7th minute and F2 from the 7th to the 14th minute of analysis). Live images of eluting cells are shown for both groups for type 1 and 2 (**B**,**D**). Cell distribution between F1 and F2 based on the calculation of the area under the curve (AUC) expressed as a percentage compared to the total area of the profile. The difference was seen between the F1 and the F2 of the 0.1 and 0.25% groups (**E**); distribution was also expressed as a number of counted cells by the software for each fraction of all samples analyzed (**F**). (t-test: $p < 0.05$ *).

For both treatments, most of the cells eluted in F1, showing a higher intensity of the peak confirmed by the area under the curve (AUC) (Figure 1E) which represents the number of eluted cells. The distribution of cells between F1 and F2 in the sample treated with a 0.25% trypsin was statistically different compared to 0.1% samples: 0.25% samples generated a higher AUC in F1 and a lower AUC for F2 compared to 0.1% samples. Post-processing analysis of counted cells confirmed the difference between F1 and F2 from both groups, even though it was no significant (Figure 1F). AECs profiling showed the ability of this technology to predict the achievement of the isolation process.

3.2. *Quality Control (QC) of Freshly Isolated AECs*

The fractionation profile gave the immediate predictivity of the AECs isolation process while the post processing analysis of the recorded data allowed a better characterization of the examined populations. The software presents the physical characteristics of each cell eluted under the camera, becoming in this way an excellent tool to increase the information besides the fingerprint obtained by the profile. Thanks to this tool, we discovered a difference between the 0.1% and 0.25% samples. When the isolation protocol was successful, AECs derived by the 0.1% showed a more heterogenous population, with two clear sub-populations F1 and F2. Scatter plot representing the cell diameter versus the time of analysis showed the two-populations distribution, F1 showed a wider dimensional distribution composed of bigger cells and cell-aggregates and a second population dimensionally more homogenous and smaller in the F2 (Figure 2A(i)). On the contrary, AECs isolated using 0.25% trypsin showed a more homogeneous population in respect of the cell dimension, with a more compact cell cloud around the diameter of 30 µm especially in F1 (Figure 2A(ii)). When cell circularity was studied, the F1 sub-population express the same average in both trypsin treatments while F2 cells of the 0.25% showed the most circular cells at the 10th minute of analysis and the 0.1% at the 13th minute, an indication of how the two concentrations differ in the membrane treatment and consequent cells release. Compared to the 0.1% AECs, 0.25% AECs are a more homogenous population. Analysis of samples that did not retrieve proliferating cells, clearly showed in the scatter plot distribution that all injected cells eluted in F1 and no cells are present at the 10th minute of the analysis which is the highest point of the second peak in type 1 profile.

Besides the scatter plot, the average of every parameter was calculated for each fraction and compared to the general population (CTRL). The results confirmed the principles of the separation process because cells with higher diameter are in F1 and smaller one elutes later in F2 (Figure 3A,D). The same trend was observed for the circularity parameter: the most circular cells are in the F2. Stem cells have the characteristics to be circular and with well-defined counters, so there is a link between this parameter and then the ability of adhesion and proliferation of these cells. Even though there is only a small difference, F2-AECs from 0.25% trypsin treatment is slightly more circular than 0.1% AECs. One of the hypotheses is that this treatment released smaller and more circular cells compared to the 0.1% treatment. When cellular aspect/ratio was measured, F2 cells showed a lower value compared to F1 cells, which is in line with the higher circularity of these cells. Despite a significant difference between F1 and F2, the difference between F1 and F2 for 0.25% samples is less highlighted than in 0.1% samples. This result is in line with the ability to isolate a more homogenous population using the 0.25% concentration.

3.3. *Viability and Cell Recovery*

Membranes treated with the two concentrations of trypsin gave very heterogenous result on cell recovery. For both the concentrations the distribution is rather wide as shown in the graph (Figure 4C). Even though there was not a significant difference, the membrane treated with a 0.1% trypsin recovered almost half of the cells of 0.25% (4.35×10^6 vs. 8×10^6 cells). At the macroscopical and microscopical observation, membrane treated with 0.1% trypsin resulted whiter and cells were still present on the membrane (Figure 4B(i,ii)) while membrane treated with the 0.25% concentration showed gel-like consistence and very few cells were seen on the membrane (Figure 4C(i,ii)). The 0.25% treatment results more efficient in cell removal as seen by a higher cell recovery (Figure 4D). For the 0.1% samples, cell recovery was equally distributed between F1 and F2 while for the 0.25% treated samples, F2 was less abundant than F1 (Figure 4E). To see the effect of the digestion process, the viability of post-fractionation cell collection was measured by counting the number of alive cells compared to the total cells in each fraction. Statistical difference was seen between F1 and F2 for the 0.1% group with the F1 have been the most vital one, while for the 0.25% group no difference was observed because of the heterogeneity of the F2 group.

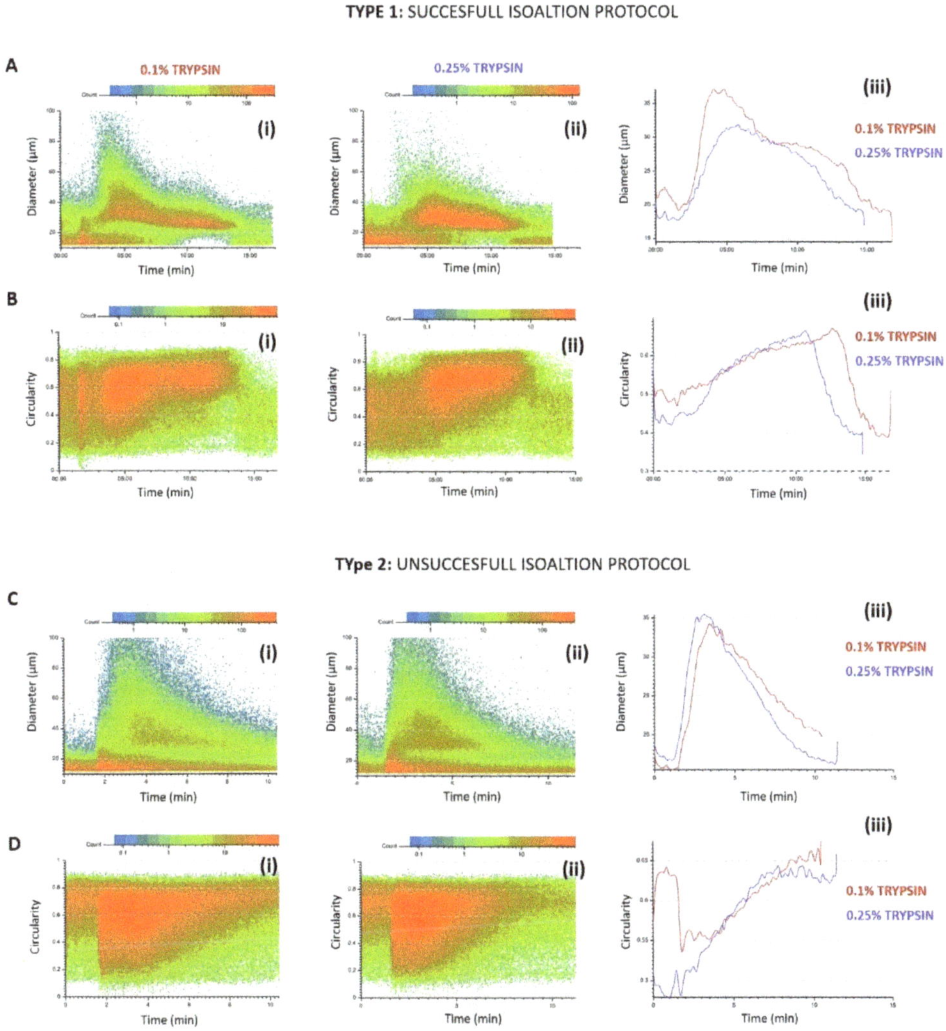

Figure 2. Representative images of the physical parameters of AECs from 0.1 and 0.25% trypsin treatments. Diameter and circularity were analyzed and compared between the two treatments. Scatter plot of diameter (**A**) and circularity (**B**) for 0.1% (**i**), 0.25% (**ii**) and the overly of the average (**iii**) for the successful protocol and for an example of an unsuccessful protocol for the cell diameter (**C**) and circularity (**D**).

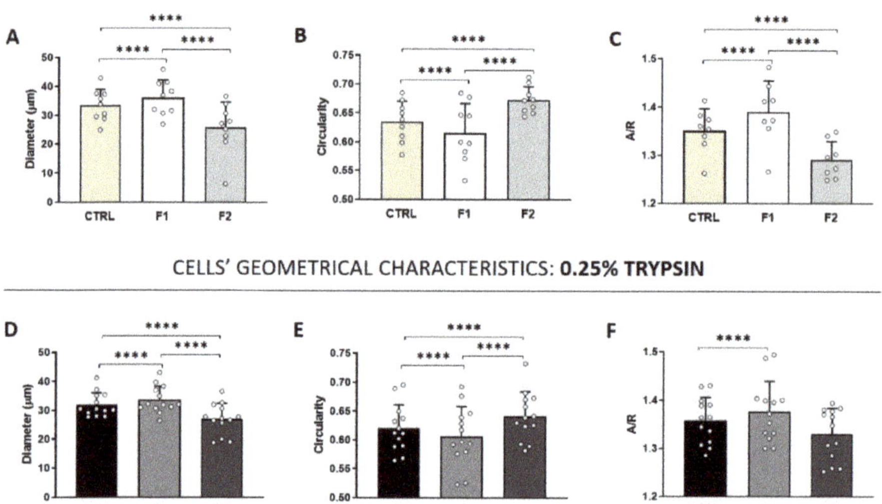

Figure 3. Cells' geometrical features for AECs derived using 0.1% trypsin (**A–C**) and 0.25% trypsin (**D–F**) for diameter, circularity and aspect/ratio (A/R). (*t*-test: $p < 0.0001$ ****).

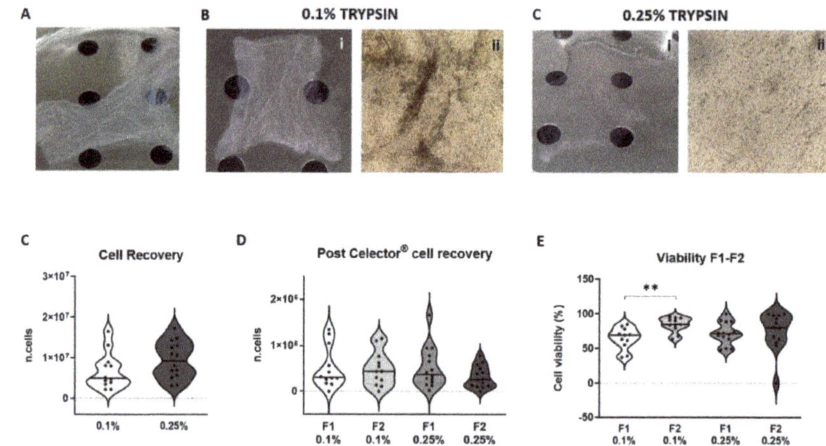

Figure 4. Representative images of amniotic membrane pre-treatment (CTRL) (**A**), macroscopic (**B-i**) and microscopic image using a 4x objective (**B-ii**) post-treatment using 0.1% trypsin and macroscopic (**C-i**) and microscopic image using a 4x objective (**C-ii**) post-treatment using 0.25% trypsin; the number of cells recovered after enzymatic treatment counted using the erythrosine solution to discriminate alive cells (**C**); the number of collected cells per analysis (run) for each fraction for both trypsin treatment, 0.1 and 0.25% (**D**); Percentage of viable cells for collected fractions F1 and F2 for both enzymatic treatment (**E**). (*t*-test: $p < 0.01$ **).

3.4. Morphology and Adhesion Properties

Morphologically, cells from both conditions had an epithelial morphology, small in size and few cells showing long pedicles (Figure 5A). In the 0.1% condition we observed few cells with wider cytoplasm compared to 0.25% which resulted homogenous and there was no difference when adherent cell area was measured (Figure 5C).

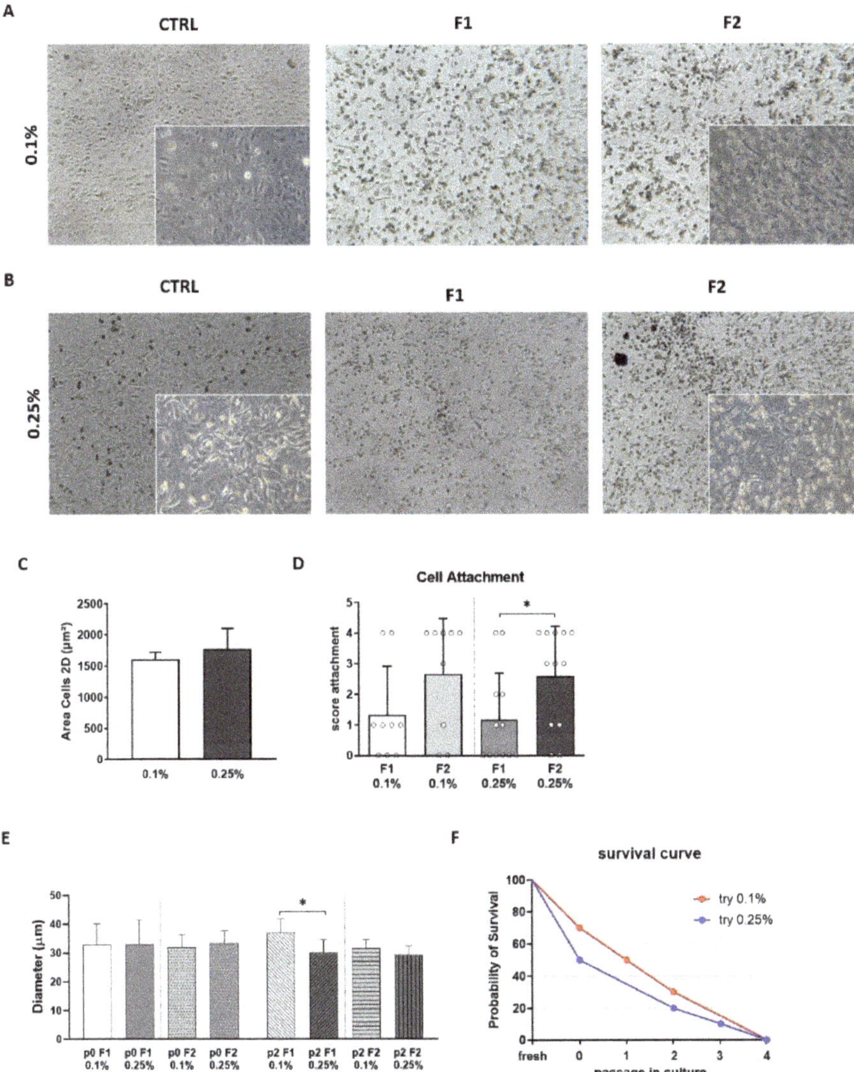

Figure 5. Representative images of AECs from both enzymatic treatment, 0.1% (**A**) and 0.25% (**B**) from control population (CTRL), F1 and F2 derived AECS and it is visible that a minority of the cells in F1 attached to the plate. AECs from 0.1% and 0.25% treatment did not show difference in area size (**C**). The ability to adhere to the plastic surface was scored (from 0 to 4, min to max) for the separated AECs from F1 and F2 for both trypsin groups. F1 derived AECs have a lower ability to adhere compared to the F2 cells, which are the ones more vital and able later to proliferate (**D**). Cell diameter was measured by post-processing analysis and cell population was divided following the same time interval used for the fresh sample. AECs at passage 0 and 2 were analyzed for both trypsin treatments (**E**). AECs control from 0.1 and 0.25% were grown in culture to monitor proliferation ability (**F**). The 0.1% AECs have a higher adhesion propensity (70%) while AECs from 0.25% adhere to the culture dish only in 50% of the cases. Even with the difference in initial adhesion, AECs grow until the 4th passage in culture. (t-test: $p < 0.05$ *).

AECs from both trypsin treatment, 0.1% and 0.25%, had the ability to adhere to plastic with a different grade, 70% for the 0.1% treated samples and 50% for the 0.25% samples. This difference could be explained by the different grade of adhesion of single fractions. For the 0.1% samples, F2 was mostly adherent and many cells from the F1, whereas in the 0.25% group there was a significant difference between F2 and F1 cells, with the latter showing a lower adhesion profile (Figure 5D). When 0.1 and 0.25% AECs were kept in culture, we saw a similar trend, meaning that the F2 component from 0.25% is probably contributing to the proliferation (Figure 5E).

Morphologically AECs from both groups have similar adhesion area and could proliferate till the 4th passage in culture. Despite the AECs in adhesion did not show any difference in the 2D cell dimension between the two concentration treatments, scatterplot graphs showed the heterogeneity of the two populations (Figure 6). AECs from 0.1% treatment showed a temporarily wider profile, with cells eluted already from the 4th to the 13th minute while 0.25% AECs eluted from the 7th till the 13th minute. The higher concentration of trypsin performs a selection on the population. To investigate even more the heterogeneity of these cells we investigated the shape and dimension of the nuclei (Figure 6). Imaging analysis of DAPI stained nuclei showed a similar distribution of normal nuclei in the 0.1% samples, with no difference between the F1 and F2 (Figure 7A). Interestingly, we observed a lower presence of normal nuclei in the F1 from 0.25% samples compared to F2 (63 vs. 77%, $p < 0.0001$). Cells derived by 0.1% trypsin treatment had a higher presence of cells with senescent nuclei (LR) compared to 0.25% cells and the majority resides in the F1 both in the 0.25 and 0.1% group (Figure 7C). Cell division was observed in culture and the 0.25% samples seemed to be more proliferating than the 0.1% even though the fraction with the higher presence of mitotic nuclei was the F1.

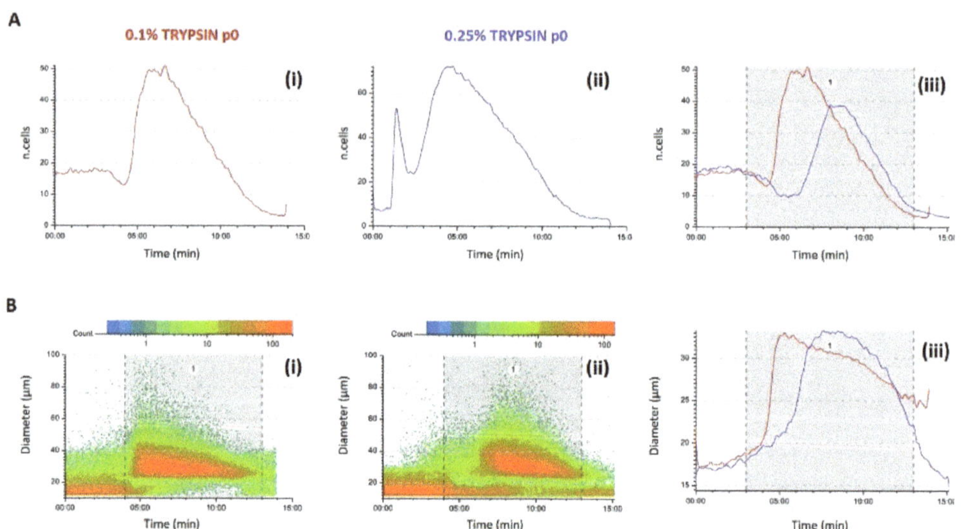

Figure 6. Representative images of the physical parameters of AECs from 0.1% and 0.25% trypsin treatments at passage 0 in culture. Profile was analyzed and compared between the two treatments (**A**). Scatter plot of diameter (**B**) for 0.1% (**i**), 0.25% (**ii**) and the overly of the average (**iii**).

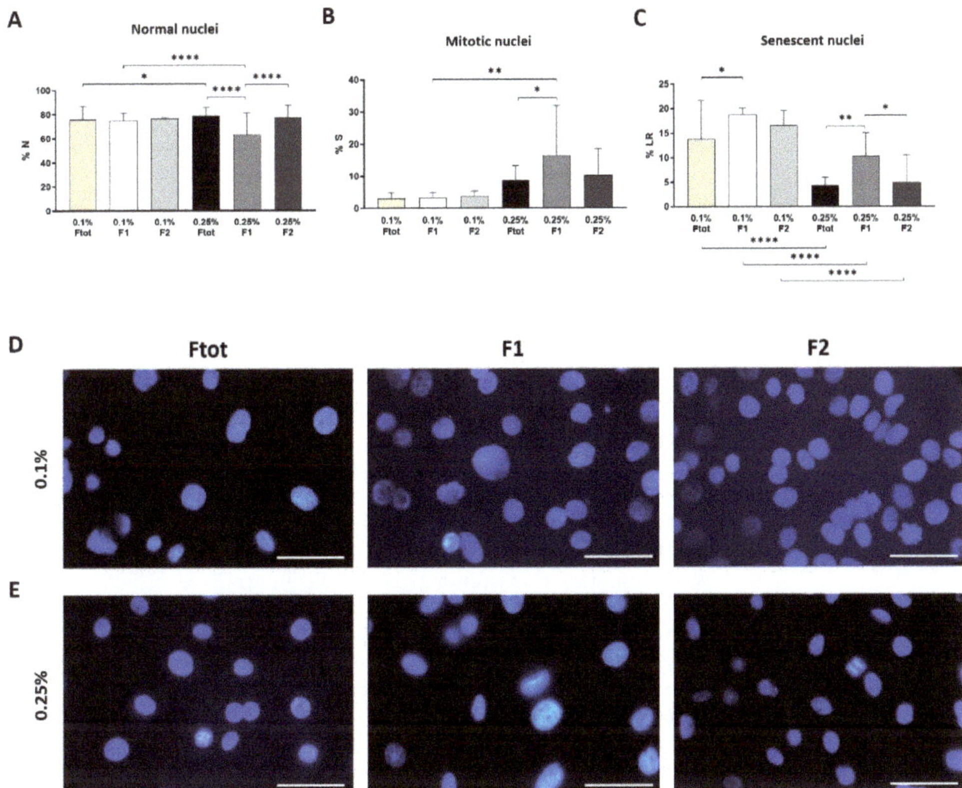

Figure 7. Morphological analysis of the nuclei showed presence of normal nuclei (**A**), mitotic (**B**) and senescent (**C**). Representative images of immunofluorescence staining for DAPI to visualize shape heterogeneity in fractions samples for both enzymatic treatments (**D**,**E**). (*t*-test. $p < 0.05$ *, $p < 0.01$ **, $p < 0.0001$ ****).

4. Discussion

This study evidenced the potential of this technology as a quality control system to predict the outcome of stem cell isolation procedure and the quality of freshly isolated cells, specifically amniotic epithelial cells (AECs).

Stem cells act following different mechanisms to treat damages: they differentiate into specialized cells, modulate host immuno-response and release specific factors that stimulate tissue regeneration. After 2018, the number of clinical trials using MSCs is slowing down because comparison among trials is difficult due to cells heterogeneity, organ origin, different cell preparation protocols and different passage numbers. Standardization of cells is necessary to obtain more reproducible and comparable results among centers [38–40]. In the last decade a huge effort has been made to ameliorate standards for cell culture and production, using optimal reagents, industrialized bioreactors where all variables such as temperature and gas exchange are constantly controlled, in order to avoid operator's variability. Despite standardize protocols and reagents may help the reproducibility of isolation protocol and the in vitro cell expansion, the inter-patient differences are still under debate. As an example, patient derived induced pluripotent stem (iPS) cells are key platform to study the impact of human cell type-specific gene regulation but inevitably, differences between donors affect most iPSC traits, from cell morphology to DNA methylation, mRNA and protein abundance to pluripotency and differentiation [41]. Even after controlling for genotype, substantial experimental heterogeneity remains.

The therapeutic potential of AECs is well known and recently, it was proven to enhances the engraftment, viability and graft function of pancreatic islet organoids in diabetes model when co-cultured, which can be used as a cell therapy approach for diabetes [19]. The possibility to use these cells in an allogenic manner raised their interest even more in the scientific community [42]. The isolation procedure is very simple consisting in the use of trypsin to detach the epithelial layer from the membrane, leaving the mesenchymal component attached to it. In our records, AECs recovery and proliferative ability were very heterogeneous and could not be linked to a macroscopical observation of the quality of the membrane, therefore we investigate the potential of the NEEGA-DF method as a quality control system. Technically, enzymatic digestion could lead to the detachment of cells in an aggregate form that unlikely adhere to the plate and could hinder single cell attachment. Clearly, cell aggregates are heavier than single cells, so through the method could be separated from the single cells. Moreover, dead and damaged cells are usually eluted in the first minute of the analysis because they do not reach the equilibrium, inferring with the quality of the cells' suspension.

Results showed the prediction of the live fractogram to define the success of the isolation protocol: bimodal profile is linked to alive, adherent and proliferating cells, whereas fractogram having one single peak or with a low intensity second peak represents the non- or poorly- adherent cells. Cells isolated using both concentrations of trypsin, 0.1 and 0.25%, showed an equal prediction of the fractogram, which means that the general action of the enzyme on the membrane is very similar. This prognostic tool can save time, reagents and effort in culturing only proliferating cells as the verification of vitality required few days because any additional movement of the dish, to observe cell adhesion at the microscope, could affect cell attachment. The QC method developed delivered the answer to the closed-question on the isolation procedure in 15-min time. Count of alive cells is not a very reliable method to define the quality of digestion, since some samples, characterized by a majority of alive cells, showed a single-peak profile and resulted in non-adherent cells. These cells are probably senescent cells unable to adhere to plastic surfaces or damaged cells by the enzymatic treatment, with a consequent selection of the solely proliferative clones. Nuclei analysis showed the presence of senescent cells mostly in F1, both in 0.1% and 0.25% samples, even though the lower concentration treatment showed a higher number of senescent nuclei both in F1 and F2 compared to 0.25% cells. We hypothesize that 0.25% concentration performs a selection, removing senescent cells, and leaving a more homogenous sample as proved by scatter plot graph representing cells' diameter after first expansion in culture. These cells elute at the same time interval of F2 proving their initial belonging of the F2 peak and the consequent no attachment of mostly F1 cells. Even at later passage, 0.25% AECs showed a more homogeneous dimensional distribution, whereas 0.1%, especially at passage 2 in culture, showed an increase in the area of cells eluted in the F1 interval. We conclude that the 0.25% treatment applied a selection on the most proliferate clones of AECs, resulting in adherent proliferating cells.

We presented for the first time an analytical platform that analyze freshly primary cells and understand the effect of isolation process on their viability and propensity to adhere to plastic. The protocol could be implemented in quality control guidelines for scale-up cell production and cryopreservation of cells for future uses with the adding value that the methods separate cells in a label-free mode, therefore no additional manipulation occurs. The possibility to avoid antibodies-sorting method, that additionally stresses cells, are requested by the scientific community [43,44] and new methodologies are valued. We developed an innovative approach of cells' physical characteristics measurements and their visualization in relation to the time of analysis using a scatter plot graph, which allowed a better understanding of the heterogeneity of complex biological samples. The separative potential of the NEEGA-DF methodology was already proven in expanded stem cells from amniotic fluid where we identified the most staminal subpopulation based only on cells profile [24] and dimensional characterization was manual and could not cover all cells analyzed. In this work, the cell imaging process was implemented and new information can be extrapolated, giving a general overview of the cell composition of the complex biological sample. Often geometrical features of cells are associated with differentiation potential and staminality, even though these studies involve 2D system and the use of special substrates. As an example, using high content imaging, Marklein et al. [45] demonstrated that nuclear morphological profiles of mesenchymal stromal cells had distinct morphological features that were highly predictive of MSC mineralization capabilities. In addition, time lapse imaging of 2D cell cultures in combination with morphological cell analysis are increasingly used to assess, track and even predict MSCs differentiation phenotypic outcomes. Dimension, circularity and aspect/ratio are the main geometrical characteristics that define cells. All these features can be obtained during the cell fractionation and related to functional ability after genetic and differentiation assessments. Hence, by converting the geometrical characteristics of sorted cells into a morphological fingerprint of the biological sample, this information may be used to improve cell culture, sorting of sub-population of interest and as we proved in this work as a prognostic marker to deduce vitality of primary cells.

5. Conclusions

In conclusion, we proved how the NEEGA-DF method and its technological implementation within Celector® technology are able to predict the outcome of the isolation of AECs using the fractogram, the fingerprint of the sample, in a very short time. Moreover, the data output of the post-processing imaging adds a new type of information of the complexity of the sample to better understand its composition and cell features for its use in therapy applications.

6. Patents

Celector® is based on a technology patented in Italy (no. IT1371772, "Method and Device to separate totipotent stem cells"), in USA, and in Canada (no. 8263359 US en. CA2649234, "Method and device to separate stem cells").

Stem Sel® has also an Italian patent (IT1426514, "Device for the Fractionation of Objects and Fractionation Method, allowed 2016).

Author Contributions: Conceptualization, S.Z. and F.A.; data curation, S.Z. and G.M.; investigation, G.M., V.P. and A.M.; methodology, S.Z., P.R., B.R. and A.Z.; resources, G.S.; supervision, F.A.; visualization, S.Z. and G.M.; writing—original draft, S.Z.; writing—review and editing, S.Z., P.R., L.B., F.A., B.R. and A.Z. All authors have read and agreed to the published version of the manuscript.

Funding: This research received no external funding.

Institutional Review Board Statement: The study was approved by the Local Research Ethics Committee (EM894-2020_68/2017/U/Tess/AOUBo).

Informed Consent Statement: Informed consent was obtained from each healthy donor before specimen collection.

Acknowledgments: We would like to thank the AGD-Associazione Giovani Diabetici (Bologna, Italy).

Conflicts of Interest: Andrea Zattoni, Barbara Roda and Pierluigi Reschiglian are associates of the academic spinoff company Stem Sel Srl (Bologna, Italy). The company mission includes the development and production of novel technologies and methodologies for the separation and characterization of cells and bio samples. All the other authors report no conflict of interest since nobody have commercial associations that might create a conflict of interest in connection with submitted manuscripts.

References

1. Cheung, T.S.; Bertolino, G.M.; Giacomini, C.; Bornhäuser, M.; Dazzi, F.; Galleu, A. Mesenchymal Stromal Cells for Graft Versus Host Disease: Mechanism-Based Biomarkers. *Front. Immunol.* **2020**, *11*, 1338. [CrossRef] [PubMed]
2. Amorin, B.; Alegretti, A.P.; Valim, V.; Pezzi, A.; Laureano, A.M.; Da Silva, M.A.L.; Wieck, A.; Silla, L. Mesenchymal stem cell therapy and acute graft-versus-host disease: A review. *Hum. Cell* **2014**, *27*, 137–150. [CrossRef]
3. Wayne, A.S.; Baird, K.; Egeler, R.M. Hematopoietic stem cell transplantation for leukemia. *Pediatr. Clin. N. Am.* **2010**, *57*, 1–25. [CrossRef] [PubMed]
4. Takami, A. Hematopoietic stem cell transplantation for acute myeloid leukemia. *Int. J. Hematol.* **2018**, *107*, 513–518. [CrossRef] [PubMed]
5. Moll, G.; Drzeniek, N.; Kamhieh-Milz, J.; Geißler, S.; Volk, H.-D.; Reinke, P. MSC Therapies for COVID-19: Importance of Patient Coagulopathy, Thromboprophylaxis, Cell Product Quality and Mode of Delivery for Treatment Safety and Efficacy. *Front. Immunol.* **2020**, *11*, 1091. [CrossRef] [PubMed]
6. Parolini, O.; Alviano, F.; Bagnara, G.P.; Bilic, G.; Bühring, H.-J.; Evangelista, M.; Hennerbichler, S.; Liu, B.; Magatti, M.; Mao, N.; et al. Concise review: Isolation and characterization of cells from human term placenta: Outcome of the first international Workshop on Placenta Derived Stem Cells. *Stem Cells* **2008**, *26*, 300–311. [CrossRef] [PubMed]
7. Vanover, M.; Wang, A.; Farmer, D. Potential clinical applications of placental stem cells for use in fetal therapy of birth defects. *Placenta* **2017**, *59*, 107–112. [CrossRef]
8. Morandi, F.; Marimpietri, D.; Görgens, A.; Gallo, A.; Srinivasan, R.C.; El-Andaloussi, S.; Gramignoli, R. Human Amnion Epithelial Cells Impair T Cell Proliferation: The Role of HLA-G and HLA-E Molecules. *Cells* **2020**, *9*, 2123. [CrossRef] [PubMed]
9. Gramignoli, R.; Srinivasan, R.C.; Kannisto, K.; Strom, S.C. Isolation of Human Amnion Epithelial Cells According to Current Good Manufacturing Procedures. *Curr. Protoc. Stem Cell Biol.* **2016**, *37*, 1E.10.1–1E.10.13. [CrossRef]
10. Miki, T.; Lehmann, T.; Cai, H.; Stolz, D.B.; Strom, S.C. Stem cell characteristics of amniotic epithelial cells. *Stem Cells* **2005**, *23*, 1549–1559. [CrossRef] [PubMed]
11. Motedayyen, H.; Zarnani, A.-H.; Tajik, N.; Ghotloo, S.; Rezaei, A. Immunomodulatory effects of human amniotic epithelial cells on naive CD4(+) T cells from women with unexplained recurrent spontaneous abortion. *Placenta* **2018**, *71*, 31–40. [CrossRef]
12. Strom, S.C.; Skvorak, K.; Gramignoli, R.; Marongiu, F.; Miki, T. Translation of amnion stem cells to the clinic. *Stem Cells Dev.* **2013**, *22* (Suppl. 1), 96–102. [CrossRef] [PubMed]
13. Okere, B.; Alviano, F.; Costa, R.; Quaglino, D.; Ricci, F.; Dominici, M.; Paolucci, P.; Bonsi, L.; Iughetti, L. In vitro differentiation of human amniotic epithelial cells into insulin-producing 3D spheroids. *Int. J. Immunopathol. Pharmacol.* **2015**, *28*, 390–402. [CrossRef]
14. Xu, H.; Zhang, J.; Tsang, K.S.; Yang, H.; Gao, W.-Q. Therapeutic Potential of Human Amniotic Epithelial Cells on Injuries and Disorders in the Central Nervous System. *Stem Cells Int.* **2019**, *2019*, 5432301. [CrossRef] [PubMed]
15. Kim, K.Y.; Suh, Y.-H.; Chang, K.-A. Therapeutic Effects of Human Amniotic Epithelial Stem Cells in a Transgenic Mouse Model of Alzheimer's Disease. *Int. J. Mol. Sci.* **2020**, *21*, 2658. [CrossRef]
16. van den Heuij, L.G.; Fraser, M.; Miller, S.L.; Jenkin, G.; Wallace, E.M.; Davidson, J.O.; Lear, C.A.; Lim, R.; Wassink, G.; Gunn, A.J.; et al. Delayed intranasal infusion of human amnion epithelial cells improves white matter maturation after asphyxia in preterm fetal sheep. *J. Cereb. Blood Flow Metab.* **2019**, *39*, 223–239. [CrossRef] [PubMed]
17. Hodges, R.J.; Lim, R.; Jenkin, G.; Wallace, E.M. Amnion epithelial cells as a candidate therapy for acute and chronic lung injury. *Stem Cells Int.* **2012**, *2012*, 709763. [CrossRef] [PubMed]
18. Andrewartha, N.; Yeoh, G. Human Amnion Epithelial Cell Therapy for Chronic Liver Disease. *Stem Cells Int.* **2019**, *2019*, 8106482. [CrossRef]
19. Lebreton, F.; Lavallard, V.; Bellofatto, K.; Bonnet, R.; Wassmer, C.H.; Perez, L.; Kalandadze, V.; Follenzi, A.; Boulvain, M.; Kerr-Conte, J.; et al. Insulin-producing organoids engineered from islet and amniotic epithelial cells to treat diabetes. *Nat. Commun.* **2019**, *10*, 4491. [CrossRef] [PubMed]
20. Liu, J.; Hua, R.; Gong, Z.; Shang, B.; Huang, Y.; Guo, L.; Liu, T.; Xue, J. Human amniotic epithelial cells inhibit CD4+ T cell activation in acute kidney injury patients by influencing the miR-101-c-Rel-IL-2 pathway. *Mol. Immunol.* **2017**, *81*, 76–84. [CrossRef] [PubMed]
21. Song, Y.-S.; Joo, H.-W.; Park, I.-H.; Shen, G.-Y.; Lee, Y.; Shin, J.H.; Kim, H.; Shin, I.-S.; Kim, K.-S. Transplanted Human Amniotic Epithelial Cells Secrete Paracrine Proangiogenic Cytokines in Rat Model of Myocardial Infarction. *Cell Transpl.* **2015**, *24*, 2055–2064. [CrossRef]
22. Zhao, B.; Liu, J.-Q.; Zheng, Z.; Zhang, J.; Wang, S.-Y.; Han, S.-C.; Zhou, Q.; Guan, H.; Li, C.; Su, L.-L.; et al. Human amniotic epithelial stem cells promote wound healing by facilitating migration and proliferation of keratinocytes via ERK, JNK and AKT signaling pathways. *Cell Tissue Res.* **2016**, *365*, 85–99. [CrossRef] [PubMed]
23. Weidinger, A.; Poženel, L.; Wolbank, S.; Banerjee, A. Sub-Regional Differences of the Human Amniotic Membrane and Their Potential Impact on Tissue Regeneration Application. *Front. Bioeng. Biotechnol.* **2020**, *8*, 613804. [CrossRef] [PubMed]
24. Centurione, L.; Passaretta, F.; Centurione, M.A.; De Munari, S.; Vertua, E.; Silini, A.; Liberati, M.; Parolini, O.; Di Pietro, R. Mapping of the Human Placenta: Experimental Evidence of Amniotic Epithelial Cell Heterogeneity. *Cell Transpl.* **2018**, *27*, 12–22. [CrossRef] [PubMed]

25. Srinivasan, R.C.; Strom, S.C.; Gramignoli, R. Effects of Cryogenic Storage on Human Amnion Epithelial Cells. *Cells* **2020**, *9*, 1696. [CrossRef] [PubMed]
26. Bilic, G.; Zeisberger, S.M.; Mallik, A.S.; Zimmermann, R.; Zisch, A.H. Comparative characterization of cultured human term amnion epithelial and mesenchymal stromal cells for application in cell therapy. *Cell Transpl.* **2008**, *17*, 955–968. [CrossRef] [PubMed]
27. Ghamari, S.-H.; Abbasi-Kangevari, M.; Tayebi, T.; Bahrami, S.; Niknejad, H. The Bottlenecks in Translating Placenta-Derived Amniotic Epithelial and Mesenchymal Stromal Cells into the Clinic: Current Discrepancies in Marker Reports. *Front. Bioeng. Biotechnol.* **2020**, *8*, 180. [CrossRef]
28. Shen, Z.-Y.; Xu, L.-Y.; Li, E.-M.; Zhuang, B.-R.; Lu, X.-F.; Shen, J.; Wu, X.-Y.; Li, Q.-S.; Lin, Y.-J.; Chen, Y.-W.; et al. Autophagy and endocytosis in the amnion. *J. Struct. Biol.* **2008**, *162*, 197–204. [CrossRef]
29. Motedayyen, H.; Esmaeil, N.; Tajik, N.; Khadem, F.; Ghotloo, S.; Khani, B.; Rezaei, A. Method and key points for isolation of human amniotic epithelial cells with high yield, viability and purity. *BMC Res. Notes* **2017**, *10*, 552. [CrossRef] [PubMed]
30. Koike, C.; Zhou, K.; Takeda, Y.; Fathy, M.; Okabe, M.; Yoshida, T.; Nakamura, Y.; Kato, Y.; Nikaido, T. Characterization of amniotic stem cells. *Cell. Reprogram.* **2014**, *16*, 298–305. [CrossRef]
31. Cui, P.; Wang, S. Application of microfluidic chip technology in pharmaceutical analysis: A review. *J. Pharm. Anal.* **2019**, *9*, 238–247. [CrossRef] [PubMed]
32. Reschiglian, P.; Zattoni, A.; Roda, B.; Casolari, S.; Moon, M.H.; Lee, J.; Jung, J.; Rodmalm, K.; Cenacchi, G. Bacteria sorting by field-flow fractionation. Application to whole-cell Escherichia coil vaccine strains. *Anal. Chem.* **2002**, *74*, 4895–4904. [CrossRef]
33. Wang, X.-B.; Yang, J.; Huang, Y.; Vykoukal, J.; Becker, F.F.; Gascoyne, P.R.C. Cell separation by dielectrophoretic field-flow-fractionation. *Anal. Chem.* **2000**, *72*, 832–839. [CrossRef]
34. Roda, B.; Lanzoni, G.; Alviano, F.; Zattoni, A.; Costa, R.; Di Carlo, A.; Marchionni, C.; Franchina, M.; Ricci, F.; Tazzari, P.L.; et al. A novel stem cell tag-less sorting method. *Stem Cell Rev. Rep.* **2009**, *5*, 420–427. [CrossRef] [PubMed]
35. Roda, B.; Reschiglian, P.; Alviano, F.; Lanzoni, G.; Bagnara, G.P.; Ricci, F.; Buzzi, M.; Tazzari, P.L.; Pagliaro, P.; Michelini, E.; et al. Gravitational field-flow fractionation of human hemopoietic stem cells. *J. Chromatogr. A* **2009**, *1216*, 9081–9087. [CrossRef]
36. Ilancheran, S.; Michalska, A.; Peh, G.; Wallace, E.; Pera, M.; Manuelpillai, U. Stem cells derived from human fetal membranes display multilineage differentiation potential. *Biol. Reprod.* **2007**, *77*, 577–588. [CrossRef] [PubMed]
37. Filippi-Chiela, E.C.; Oliveira, M.M.; Jurkovski, B.; Jacques, S.M.C.; da Silva, V.D.; Lenz, G. Nuclear morphometric analysis (NMA): Screening of senescence, apoptosis and nuclear irregularities. *PLoS ONE* **2012**, *7*, e42522. [CrossRef] [PubMed]
38. Kabat, M.; Bobkov, I.; Kumar, S.; Grumet, M. Trends in mesenchymal stem cell clinical trials 2004–2018: Is efficacy optimal in a narrow dose range? *Stem Cells Transl. Med.* **2020**, *9*, 17–27. [CrossRef] [PubMed]
39. Pamies, D.; Bal-Price, A.; Simeonov, A.; Tagle, D.; Allen, D.; Gerhold, D.; Yin, D.; Pistollato, F.; Inutsuka, T.; Sullivan, K.; et al. Good Cell Culture Practice for stem cells and stem-cell-derived models. *ALTEX* **2017**, *34*, 95–132. [CrossRef] [PubMed]
40. Pamies, D.; Bal-Price, A.; Chesne, C.; Coecke, S.; Dinnyes, A.; Eskes, C.; Grillari, R.; Gstraunthaler, G.; Hartung, T.; Jennings, P.; et al. Advanced Good Cell Culture Practice for human primary, stem cell-derived and organoid models as well as microphysiological systems. *ALTEX* **2018**, *35*, 353–378. [CrossRef]
41. Kilpinen, H.; Goncalves, A.; Leha, A.; Afzal, V.; Alasoo, K.; Ashford, S.; Bala, S.; Bensaddek, D.; Casale, F.P.; Culley, O.; et al. Common genetic variation drives molecular heterogeneity in human iPSCs. *Nature* **2017**, *546*, 370–375. [CrossRef] [PubMed]
42. Zhang, Q.; Lai, D. Application of human amniotic epithelial cells in regenerative medicine: A systematic review. *Stem Cell Res. Ther.* **2020**, *11*, 439. [CrossRef] [PubMed]
43. Pfister, G.; Toor, S.M.; Nair, V.S.; Elkord, E. An evaluation of sorter induced cell stress (SICS) on peripheral blood mononuclear cells (PBMCs) after different sort conditions–Are your sorted cells getting SICS? *J. Immunol. Methods* **2020**, *487*, 112902. [CrossRef] [PubMed]
44. Binek, A.; Rojo, D.; Godzien, J.; Rupérez, F.J.; Nuñez, V.; Jorge, I.; Ricote, M.; Vázquez, J.; Barbas, C. Flow Cytometry Has a Significant Impact on the Cellular Metabolome. *J. Proteome Res.* **2019**, *18*, 169–181. [CrossRef] [PubMed]
45. Marklein, R.A.; Surdo, J.L.L.; Bellayr, I.H.; Godil, S.A.; Puri, R.K.; Bauer, S.R. High Content Imaging of Early Morphological Signatures Predicts Long Term Mineralization Capacity of Human Mesenchymal Stem Cells upon Osteogenic Induction. *Stem Cells* **2016**, *34*, 935–947. [CrossRef] [PubMed]

micromachines

Article

Flexible Enzymatic Glucose Electrochemical Sensor Based on Polystyrene-Gold Electrodes

Annika Müsse [1,2,3,*], Francesco La Malfa [1,4], Virgilio Brunetti [1], Francesco Rizzi [1,*] and Massimo De Vittorio [1,4]

1 Center for Biomolecular Nanotechnologies, Istituto Italiano di Tecnologia, Via Eugenio Barsanti 14, 73010 Lecce, Italy; francesco.lamalfa@iit.it (F.L.M.); virgilio.brunetti@iit.it (V.B.); massimo.devittorio@iit.it (M.D.V.)
2 Dipartimento di Scienze e Tecnologie Biologiche e Ambientali, Università del Salento, Via per Monteroni, 73100 Lecce, Italy
3 Faculty of Technology and Bionics, Rhine-Waal University of Applied Sciences, Marie-Curie-Straße 1, 47533 Kleve, Germany
4 Dipartimento di Ingegneria dell' Innovazione, Università del Salento, Via per Monteroni, 73100 Lecce, Italy
* Correspondence: annika.muesse@iit.it (A.M.); francesco.rizzi@iit.it (F.R.)

Abstract: Metabolic disorders such as the highly prevalent disease diabetes require constant monitoring. The health status of patients is linked to glucose levels in blood, which are typically measured invasively, but can also be correlated to other body fluids such as sweat. Aiming at a reliable glucose biosensor, an enzymatic sensing layer was fabricated on flexible polystyrene foil, for which a versatile nanoimprinting process for microfluidics was presented. For the sensing layer, a gold electrode was modified with a cysteine layer and glutaraldehyde cross-linker for enzyme conformal immobilization. Chronoamperometric measurements were conducted in PBS buffered glucose solution at two potentials (0.65 V and 0.7 V) and demonstrated a linear range between 0.025 mM to 2mM and an operational range of 0.025 mM to 25 mM. The sensitivity was calculated as 1.76μA/mM/cm^2 and the limit of detection (LOD) was calculated as 0.055 mM at 0.7 V. An apparent Michaelis–Menten constant of 3.34 mM (0.7 V) and 0.445 mM (0.65 V) was computed. The wide operational range allows the application for point-of-care testing for a variety of body fluids. Yet, the linear range and low LOD make this biosensor especially suitable for non-invasive sweat sensing wearables.

Keywords: glucose; glucose oxidase; amperometric biosensor; body fluids; sweat; wearable sensor

Citation: Müsse, A.; La Malfa, F.; Brunetti, V.; Rizzi, F.; De Vittorio, M. Flexible Enzymatic Glucose Electrochemical Sensor Based on Polystyrene-Gold Electrodes. *Micromachines* **2021**, *12*, 805. https://doi.org/10.3390/mi12070805

Academic Editors: Maria Serena Chiriacò, Elisabetta Primiceri and Francesco Ferrara

Received: 28 May 2021
Accepted: 5 July 2021
Published: 7 July 2021

1. Introduction

To prevent and optimally treat diseases, monitoring the health status of patients is of great importance. Among one of the most widespread diseases is diabetes, a metabolic disorder that affects more than 450 million people worldwide, which is characterized by persistent high blood glucose levels that cause vascular damage and affect the heart, eyes, kidneys and nerves. The number of affected people is expected to increase up to 700 million in 2045, and it is estimated that half of all people with diabetes are undiagnosed, which illustrates the large demand for glucose monitoring [1].

Typically, in a clinical setting, glucose levels are measured invasively using frequent blood sampling, either analyzed in central laboratories or directly at the patient's bedside, known as point-of-care testing. The great advantage of using point-of-care instruments in diabetic patients is that the turn-around time is much shorter, which is crucial as changes in the blood sugar level can require immediate adjustment. In addition, less blood volume is needed, which reduces the probability for anemia due to frequent sampling [2]. In a home setting, most commonly, finger-prick testing is used, which can be painful and does not allow for frequent measurements. These drawbacks of conventional invasive glucose testing have been addressed in recent years, and increasing efforts have been made to develop minimally- and non-invasive methods [3]. Especially, wearable devices have been in focus, as they can allow for continuous measurements [4,5]. This is possible thanks to the correlation between blood glucose levels and the level in other body fluids, such as interstitial fluid, sweat and saliva (Table 1) [6–8].

Table 1. Typical glucose concentrations in different types of human body fluids. Data from ref. [9].

Body Fluid Type		**Blood**	**Interstitial Fluid**	**Sweat**	**Saliva**
Glucose concentration	Healthy patients	4.9–6.9 mM	3.9–6.6 mM	0.06–0.11 mM	0.23–0.38 mM
	Diabetic patients	2–40 mM	1.99–22.2 mM	0.01–1 mM	0.55–1.77 mM

To date, a completely non-invasive glucose sensor still is, to the best of our knowledge, not widely available for commercial use despite the high attention this technology gained recently [3,10]. The availability of large amounts of fluid that do not require withdrawal techniques makes saliva an interesting target for glucose sensing and led to the development of such sensors [11]. Yet, despite the correlation between glucose levels in blood and saliva, measurements are only reliable under fasting conditions, which limits usability for patients with diabetes and thus commercialization [12]. On a commercial level, electrochemical biosensors to measure glucose levels in interstitial fluid have been developed [13,14]. A possibility to access interstitial fluid is to access the fluid internally using needles, which penetrate through the skin. These indwelling sensors are widely available and have gained growing market acceptance [12], but potential drawbacks such as the risk for microbial infection remain [15,16]. One such commercially available sensor is the FreeStyle Libre Flash Glucose Monitoring System (Abbott Diabetes Care) with a sensor needle length of 8.5 mm [17]. In addition, less invasive sensors using microneedles to access the fluid, such as the device from Arkal Medical, have been developed [18], but swelling and irritation can occur each time the sensor breaches the skin [12]. Withdrawing the fluid from the skin is an alternative approach that has been exploited in laboratory [19] and by the commercially approved GlucoWatch® system (Cygnus Inc., Redwood City, CA, USA) in the early 2000s [20], but the device had to be retracted from the market. Among the reasons was skin irritation due to the current that was necessary during the reverse iontophoresis process to induce fluid migration across the skin [3]. Other drawbacks of interstitial fluid sensing requiring extraction are the increased lag time and possible contamination with sweat [12].

Due to the possibility to access sweat non-invasively, this body fluid has been brought into focus, in recent years, as a good candidate to allow for non-invasive sensing, but it has been pointed out that, compared with the development of glucose sensors for interstitial fluid, the commercialization of sweat glucose sensors is still low [12]. Although sweat is easily accessible because the whole body is covered with sweat glands, high sweat rates are typically found only in people such as athletes or workers and, indeed, many sensors focus on people during physical activity to guarantee sufficient sample volumes [21–23]. To enable sweat sensing in resting people, often, induced sweating is employed, for example, by local application of sweat stimulants such as Carbachol or Pilocarpine using a reverse iontophoresis approach [24,25]. Further efforts have been made to develop sensors able to operate at low volume levels of 1–5 μL [26].

However, comparatively low glucose concentrations and low excretion rates in resting people represent challenges that have to be addressed by fabricating sensors with low detection limits and taking into account the suitability of microfluidic systems to collect such small volumes [27]. This means that, in order to be able to have a versatile substrate applicable also for low volume microfluidics, the material should show low absorption and low water vapor permeability, which could change the concentrations. Thus, biocompatible polymers with the aforementioned properties such as polystyrene (PS) are preferable over the widely used silicone-based elastomer polydimethylsiloxane (PDMS) to overcome some of these drawbacks [28]. In addition, faster fabrication processes are possible using thermoplastic polymers as PS [29,30].

Ever since the first electrochemical glucose sensor for blood was developed in 1962 by Clark and Lyons [31], an electrochemical approach is still being chosen for most glucose sensors. Enzyme-based sensors allow for high sensitivity and good reproducibility while production is usually possible in a low-cost range [9]. Different sensor generations are distinguished in literature based on the enzymatic reaction side [32,33]. Glucose oxidase (GOx) sensors are based on the enzymatically catalyzed oxidation of glucose to glucono-lactone in the presence of oxygen. The coenzyme flavin adenine dinucleotide (FAD) is required as electron acceptor in this reaction and is then regenerated by reacting with O_2 to generate hydrogen peroxide (H_2O_2) [34].

$$\text{GOx(FAD)} + \text{Glucose} \rightarrow \text{GOx(FADH}_2\text{)} + \text{Gluconolactone} \quad (1)$$

$$\text{GOx(FADH}_2\text{)} + \text{O}_2 \rightarrow \text{GOx(FAD)} + \text{H}_2\text{O}_2 \quad (2)$$

When a sufficient potential is applied at the electrodes, H_2O_2 is oxidized, and a current can be measured which correlates with the amount of H_2O_2 that has been produced and thus correlates indirectly with the glucose concentration present in the fluid.

$$\text{H}_2\text{O}_2 \rightarrow \text{O}_2 + 2\text{H}^+ + 2\text{e}^- \quad (3)$$

The aim of the study was to obtain a simple enzymatic glucose sensor with a range suitable for sweat glucose sensing (see Table 1) to be integrated in a microfluidics in order to obtain a wearable device with an efficient sweat collection. The sweat sensors analyzed in Table 2 usually show the sweat collecting system being fabricated on top of the functionalized electrode. Here, a fabrication process for microfluidic systems was adapted to suit a PS nanoimprinting process that allows for a versatile, scalable and cost-effective fabrication in the context for wearables as well as point-of-care testing. A combination of a nanoimprinting lithography of a microfluidic on a PS substrate followed by electrode definition by metal evaporation enables a route for a mass production of wearable sweat sensors preserving the mechanical and electrical integrity of the electrodes. This approach can be worthwhile for envisioning a fast roll-to-roll production of non-invasive wearable sensing systems for sweat and other biological fluids.

Table 2. Comparison of present work to other amperometric glucose biosensors.

Operational Range [mM]	Limit of Detection [mM]	Sensitivity	Ref.
0.025–25	0.055	1.76 μA/mM/ cm^2	Present work
1.5–7	0.94	2.65 μA/mM/ cm^2	[35]
0.01–0.7	0.01	1 μA/mM	[36]
0–0.1	-	2.1 μA/mM	[25]
0–0.2	-	2.35 nA/μM	[22]
0.005–1	-	-	[11]
0.05–0. 2	-	3.29 nA/μM	[21]
2–10	0.05		[23]
0.005–2.8	0.005		[26]
0–0.1	0.003	23 nA/μM	[19]

2. Materials and Methods

2.1. Materials and Reagents

For the electrode fabrication, polystyrene (PS) foil with a thickness of 0.19 mm was purchased from GoodFellow (Prodotti, Gianni S.r.l., Milan, Italy) and adhesive foil sheets were obtained from Greiner Bio-one (platesealer EASYseal^{TM} transparent, RS Components S.r.l., Milan, Italy). Isopropyl alcohol (IPA), L-cysteine (BioUltra, ≥98.5% (RT)), phosphate buffered saline tablets (PBS), Bovine serum albumin (BSA, lyophilized powder, ≥96%), glycerol (≥99%), glutaraldehyde (GTA, Grade I, 70% in H_2O), glucose oxidase (GOx, from *Aspergillus niger*, Type X-S, lyophilized powder, 100,000–250,000 units/g), D-(+)-Glucose (≥99.5%) and potassium hexacyanoferrate(II) trihydrate (98.5–102.0%), were provided by Sigma Aldrich (Merck Life Science S.r.l, Milan, Italy). Deionized (DI) water was taken from a Milli-Q® water system (Millipore).

For fabrication of the microfluidics, the following material was used: Si wafer, polyester film photomasks (JD Photo Data, Hitchin, UK), SU-8 2002 and SU-8 2100 photoresist, SU-8 developer (MicroChem Corp, Newton, MA, USA), UV-glue (NOA 68, Norland Products Inc, Cranbury, NJ, USA).

2.2. Electrode Fabrication

For the experiments to characterize the properties of the working electrodes (WE), these were fabricated as single electrodes on PS foil, and measurements were taken in a electrochemical cell with a separate silver/silver-chloride (Ag/AgCl) reference electrode and a separate platinum (Pt) counter electrode; whereas for the first experiments for integration with microfluidics, the WE was combined with a counter electrode (CE) and a reference electrode (RE) on the same PS foil substrate. In all three electrodes, a 5 × 8 mm contact pad and a 10 × 0.5 mm wire connection were present, leading to the electrode surface in contact with the fluid. The circular shaped WE had a diameter of 4 mm, and, in the three-electrode system on PS foil, it was centered between the CE and RE; both CE and RE were bent in half-circular shape surrounding the WE in order to allow for the CE and RE to be close to the WE, exploiting a large surface area. The PS foil was cleaned using IPA and DI water and then dried using nitrogen flux. Using a laser cutter (VLS2.30DT, Universal Laser Systems GmbH, Wien, Austria), the design was cut into adhesive foil sheets that were attached on the PS foil to serve as mask in the following metal evaporation process. First, a thin adhesion layer of chromium (about 15 nm) was thermally evaporated on the PS film followed by a gold (Au) layer (about 150 nm). Then, the adhesive mask was carefully removed and excess PS film was cut. For the RE, Ag (about 100 nm) was thermally evaporated using a new mask. The maximum deposition rates were up to 1 Å/s.

2.3. Electrode Functionalization with GOx

Prior to the functionalization of the WE electrode, the samples were washed with IPA and DI water. Then, the electrode was covered with 50 mM L-cysteine solution for 20 h at room temperature (RT) to create thiol–gold bonds, followed by washing with DI water and BSA solution (30 mg/mL BSA in PBS). After that, 30 µL drops of GTA solution (2.5 wt% GTA, 50 mg/mL BSA and 1 vol% glycerol) were applied on the electrode surface for 2–3 h at RT to immobilize GOx via cross-linking. BSA and glycerol contributed to the stabilization. The electrodes were then washed with BSA solution to which glycerol was added (30 mg/mL BSA and 1 vol% glycerol in PBS). Next, 30 µL drops of GOx solution (25 mg/mL GOx in PBS) were placed on the electrode for 2.5 h at 4 °C. Finally, electrodes were washed with PBS solution and stored at 4 °C in PBS.

2.4. Electrode Characterization

For the electrochemical characterization of the WE electrode, experiments were conducted in an electrochemical cell with external RE (Ag/AgCl in KCl) and external CE (Pt sheet) using a potentiostat (Autolab, Metrohm Autolab, The Netherlands) and the software NOVA (Metrohm Autolab). Cyclic voltammetry (CV) was performed on Au electrodes without functionalization from −0.2 V to 0.6 V at different scan rates (10, 20, 40, 50, 60, 80, 100, 140, 180, 200 mV/s) in $[Fe(CN)6]^{3-/4-}$ (ferro-ferricyanide) solution for confirmation of the response of the electrode. Further, Au electrodes were tested for their current response in different concentrations of H_2O_2 (0–25 mM) at 0.7 V. Chronoamperometry (CA) was performed by placing the GOx-functionalized WE, the Ag/AgCl RE and the Pt CE in the electrochemical cell filled with 20 mL of PBS. After a stable baseline current was reached, glucose stock solution (2 M in PBS, prepared the previous day to allow for mutarotation of the glucose) was added stepwise to the PBS solution until a maximum glucose concentration of 25 mM was reached. The applied potentials were 0.65 V and 0.7 V at a constant pH of 7.4, which is the standard physiological buffer. To test for a possible influence of the pH, the current response was measured for pH values between pH 4.5 to pH 8 in glucose solution (pH adjusted in 1 mM glucose in PBS) and at potentials ranging from 0.5 V to 0.75 V. Sensitivity was calculated as the slope for the linear range divided by the circular electrode area, and the limit of detection (LOD) was calculated as three times the standard deviation of the baseline current divided by the slope [37]. The apparent Michealis–Menten constant Km(app) was calculated using the software OriginPro 2018 (OriginLAB, USA) following the Lineweaver–Burk formula:

$$1/I_{SS} = 1/I_{max} + (Km(app))/I_{max}) \times 1/c, \quad (4)$$

where c represents glucose concentration, I_{SS} is the steady-state current at a certain glucose concentration and I_{max} describes the maximum current under saturated conditions.

2.5. Microfluidics Fabrication for Further Device Integration

A nanoimprinting approach was used to fabricate microfluidics in PS in order to be integrated with the electrodes. Photomasks for the fabrication of a SU-8 stamp were designed using the software CleWin (WieWeb software, Hengelo, The Netherlands) and printed on photomask foil. To fabricate the stamp, a photolithography process was exploited: After oxygen plasma treatment for surface activation (100 W for 5 min), a 2 µm layer of SU-8 2002 photoresist was spin coated on a clean 2″ Si-wafer substrate after the following protocol: 500 rpm for 5 s, then 1800 rpm for 30 s; followed by a soft bake at 95 °C for 1 min 30 s; cool down period of 10 min; flood exposure under UV light (1 min 30 s at about 6 mW/cm^2 at 365 nm wavelength); post-exposure bake at 95 °C for 1 min 30 s; development for 1 min. The purpose of this thin SU-8 layer was to improve the attachment of the following thicker SU-8 layer during the imprinting process. In a second step, a layer of 110 µm of SU-8 2100 was spin coated: 1200 rpm for 90 s; soft bake for 5 min at 65 °C, then 45 min at 95 °C; UV exposition with 250 mJ/cm^2 using the photomask; post-exposure bake for 5 min at 65 °C, then 30 min at 95 °C; development 5 min; hard bake for 10 min at 150 °C. The obtained SU-8 structure is the negative of the desired microfluidic pattern to be transferred to the polymer substrate. The PS foil was rinsed with DI water and dried using nitrogen flux prior to the nanoimprinting process. The SU-8 stamp and the PS foil were stacked, and the pattern was imprinted at a temperature of 140 °C for 300 s using the nanoimprinting instrument (EITRE 3, Obducat). The channel height was measured using a profilometer (Bruker Dektat XT) resulting in a final value of 105 µm. At this stage, the electrode definition is conducted by metal evaporation with the same procedure as described in Section 2.2. The PS sample was then exposed to oxygen plasma at 200 W for 10 min (RFG 300, Diener) to hydrophilize the surface. A closed microfluidic system was obtained by applying UV glue at the borders of the microfluidics and irradiated for 1 to 1 min 30 s.

3. Results

3.1. Electrochemical Characterization

CV was performed on the Au electrodes in ferro-ferricyanide solution to characterize the current response to evaluate the electrode fabrication process for its suitability and reliability. Figure 1a shows an exemplary CV with both oxidation and reduction peak demonstrating the reversibility of the redox reaction. In addition, Figure 1b shows a plot of the square root of the scan rate against the maximum oxidation and reduction currents. The second plot showed increasing absolute current values for higher scan rates, which was expected for Au electrodes in ferro-ferricyanide solution. The characterization showed a good repeatability among several samples; thus, the fabricated Au electrodes were suitable for further functionalization steps.

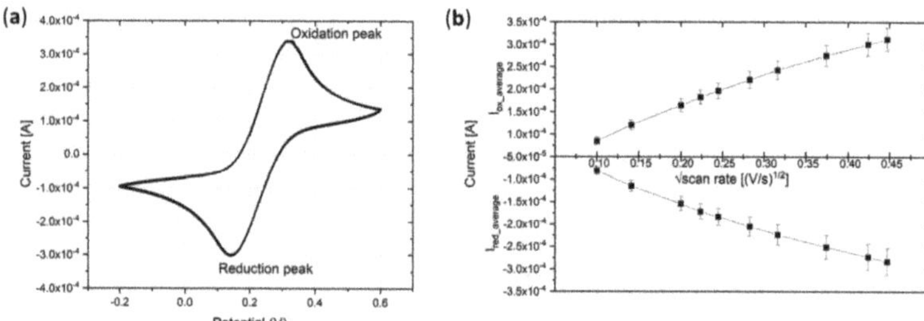

Figure 1. Cyclic voltammetry of a Au electrode on polystyrene (PS) foil in ferro-ferricyanide solution. (**a**) Exemplary cyclic voltammogram. (**b**) Square root of the scan rate vs. maximum absolute values of reduction and oxidation currents ($n = 6$). n = number of analyzed samples.

The ability of the Au electrode to detect current changes in different H_2O_2 concentrations was demonstrated as the measured current increased with increasing H_2O_2 concentrations (Figure 2). This is of importance as the sensor working principle relies on a current response to the H_2O_2 oxidation process that occurs as a result of the enzymatic glucose catalysis. As the chosen potential of 0.7 V vs. Ag/AgCl resulted in this reliable current response, further experiments were conducted at this potential.

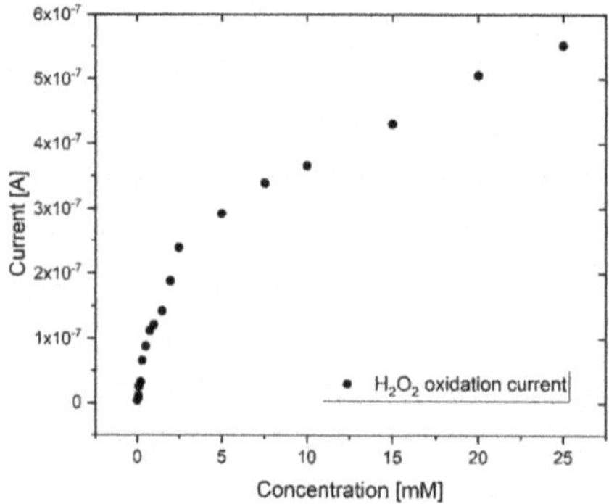

Figure 2. Exemplary current response of a Au electrode in different H_2O_2 concentrations. The trend showed an increasing current response with increasing H_2O_2 concentrations.

Glucose sensing was achieved by the functionalization of the Au electrode with GOx, where thiol–gold bonds, thanks to the cysteine self-assembled monolayer (SAM layer), acted as a link between the enzyme and the Au surface, and the GTA/BSA/glycerol network enhanced immobilization and stability (Figure 3). Chronoamperometric measurements were conducted at 0.65 V and 0.7 V to demonstrate suitability for glucose sensing. In addition to an applied potential of 0.7 V, a slightly lower value of 0.65 V was chosen to test for the sensor performance, as lower potentials can be advantageous to reduce interference with other species [38]. A linear increase in the current was observed between concentrations of 0.025 mM and 2 mM at 0.7 V, whereas the whole operational range was observed to be between 0.025 mM and 25 mM of glucose for both 0.65 V and 0.7 V, as saturation occurred at concentrations higher than 25 mM (see Figure 3). Comparing the current between an applied potential of 0.65 V and 0.7 V, it is notable that the overall current is higher for 0.7 V. The sensitivity was calculated as 1.76 $\mu A/mM/cm^2$ at 0.7 V. The LOD was calculated to be 0.055 mM at 0.7 V. The linear range obtained at a potential of 0.7 V made the higher potential seem more favorable. However, it has to be noted that, at a higher potential, the baseline current was higher and showed higher noise levels, and a 5–10 min longer period of time was required at 0.7 V potential before a stable baseline current was obtained. The apparent Michaelis–Menten constant was computed as 0.445 mM for 0.65 V and 3.34 mM for 0.7 V, which showed that GOx had a higher substrate affinity at the lower potential. However, the lower substrate affinity at 0.7 V showed better suitability for the determination of glucose concentrations typically present in sweat and other body fluids. As shown in Figure 3b, at 0.7 V the Michaelis–Menten curve shows a saturation at higher concentrations which implicates that, even at higher glucose concentrations, the enzyme activity is not limiting the reaction rate.

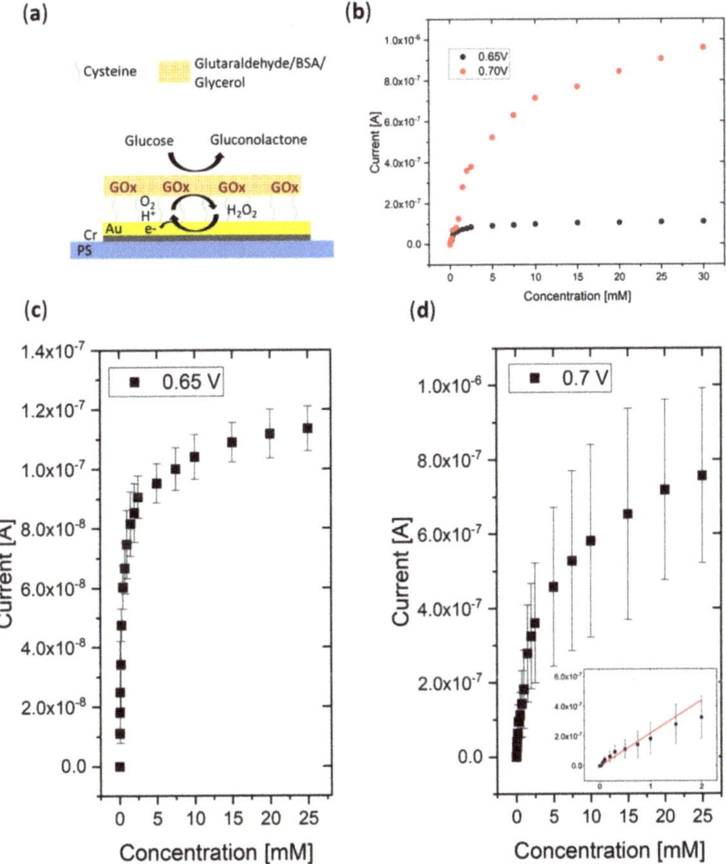

Figure 3. (**a**) Scheme of sensing mechanism of the glucose oxidase (GOx)-functionalized Au electrode. (**b**) Exemplary calibration plots of the GOx-functionalized Au electrode in different glucose concentrations for (**c**) 0.65 V (n = 3) and (**d**) 0.7 V (n = 3). A linear increasing current response for increasing glucose concentrations was found between 0.025 mM and 2 mM at 0.7 V (inset), whereas the operational range was up to 25 mM at both potentials before saturation occurred. n = number of analyzed samples.

To investigate the relation between the pH of the analyzed fluid and the response of the electrode, the current was measured at pH values between 4.5 and 8 for potentials between 0.5 V and 0.75 V (Figure 4). At a constant glucose concentration of 1 mM, which was chosen because this concentration is well within the linear range, the current was found to increase with increasing pH values despite the standard deviations overlapping in adjacent pH values mainly between pH 5 and pH 6. Moreover, an overall trend was found, which showed increasing current with increasing potential.

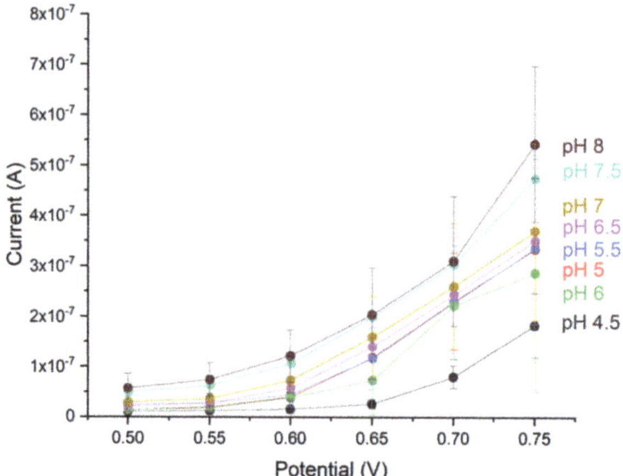

Figure 4. Current response of GOx functionalized Au electrode at different pH values (n = 3). n = number of analyzed samples.

3.2. *Microfluidics for Three-Electrode System*

The nanolithography and nanoimprinting process used for the fabrication of a simple microfluidic system with inlet, outlet and a circular chamber is described in Figure 5a. Following the design scheme of the three-electrode system and the microchannel (Figure 5b), the Au WE and CE and the Ag RE were evaporated on flexible PS foil (Figure 5c) and a closed microfluidic system was tested for the fluid dispersal using dyed water, which showed uniform fluid flow over all electrodes (Figure 5d).

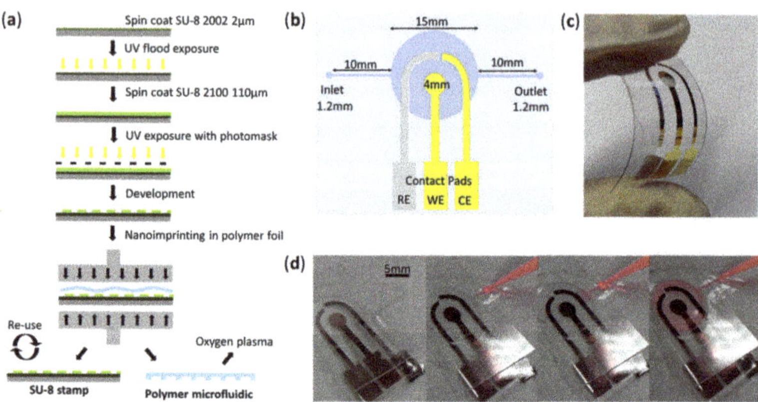

Figure 5. (**a**) Microfluidic fabrication process using nanoimprinting. (**b**) Scheme of the three-electrode system and a simple microfluidic channel. (**c**) Reference electrode (RE), working electrode (WE) and counter electrode (CE) evaporated on PS foil. (**d**) Application of dyed water to validate the filling of the chamber.

4. Discussion

A first generation amperometric biosensor for glucose was realized by immobilizing GOx on an Au electrode that was evaporated on flexible PS polymer foil. The enzyme was attached to the Au surface via thiol-bonds created by a cysteine SAM layer, and embedded into the cross-linker GTA. BSA and glycerol were added for stabilization purposes.

The response to different glucose concentrations was evaluated by chronoamperometric measurements, and a good sensor output for the biologically relevant range in body fluids such as blood, interstitial fluid, sweat and saliva was demonstrated. Comparing the sensor performance at 0.65 V and at 0.7 V showed that an LOD of 0.055 mM glucose and sensitivity of 1.76 μA/mM/cm^2 were obtained for the higher potential, whereas, at 0.65 V, no linear range was shown. A higher substrate affinity was found at 0.65 V given by the apparent Michaelis–Menten constant of 0.445 mM compared with 3.34 mM at 0.7 V. The low Km(app) value at 0.65 V compared with a similar sensor fabricated with GTA for cross-linking (Km(app) of 1.15 mM [35]), could be due to the stabilizing effect of BSA and glycerol. In general, the obtained results for the sensor are in line with the previous work based on a GOx sensor; however, a comparatively large operational range was obtained (Table 2), also possibly due to the addition of BSA and glycerol as stabilizing agents [39]. To reduce possible interference effects in body fluids, an electrostatically charged and porous membrane such as a nafion layer could be added [40]. Further testing and assessing the optimal potential can help to find the best trade-off between the sensor characteristics depending on the requirements for the field of application.

The linear range of 0.025 mM to 2 mM and the large operational range of 0.025 mM to 25 mM make the sensor suitable for sensing glucose concentrations in a variety of body fluids (see Table 1). A possible field of application of the sensor is its use as a disposable strip for point-of-care measurements to detect the glucose concentration, which is to be used by medical professionals. In this context, the whole three-electrode system can be placed on the same strip, so the presented electrode fabrication process demonstrated the first steps for further integration. Wearable device market applications could be envisioned, especially sweat sensors that allow for non-invasive glucose monitoring.

Several adaptations should be considered for a reliable sweat sensor. In fact, it is known that sweat pH can vary between 4.5 and 7 [41], and the results showed that the current varies depending on the pH. Therefore, it could be of advantage to integrate a pH sensor to achieve a more complete calibration. In addition, it is noteworthy that a response maximum was expected at a slightly acidic pH, as the pH optimum of most GOx is between a pH of 5 and 6 [42]. The different behavior of GOx on the sensor surface could be due to nonspecific modifications of the enzyme surface during the fabrication process [43], and further work is necessary to gain full understanding. Another important adaptation regards the RE: When fabricating a reliable RE for sweat, the presence of chloride ions needs to be considered. In case of the Ag/AgCl RE, an additional layer is required to avoid sensing the presence of chloride ions. Such an electrode can be realized by chemically converting the evaporated Ag electrode into an Ag/AgCl electrode and subsequently adding a layer of polyvinyl butyral [44]. As secreted sweat volumes range between 0.1 and 2 μL/min/cm^2 [12], the miniaturization of the electrodes will be of advantage to collect and drive such small amounts of fluid. On that account, the presented fabrication technique for the microfluidics is highly versatile and allows to easily adapt to smaller structure sizes. Even other thermoplastic materials can be used for the nanoimprinting process, among these, more flexible polymers such as soft thermoplastic elastomers and specifically styrenic block copolymers that can adapt well to the human skin because their Young's modulus is more similar to that of skin [45]. Long-term-stability and testing of real body fluid samples will be of importance during the development of a wearable sweat sensing device that allows for continuous glucose measurements and, by this, paving the way for a new generation of non-invasive glucose sensors improving the quality of life of diabetic patients.

Author Contributions: Conceptualization, A.M., F.L.M., V.B., F.R. and M.D.V.; methodology, A.M., F.L.M. and V.B.; formal analysis, A.M. and F.L.M.; investigation, A.M. and F.L.M.; data curation, A.M. and F.L.M.; writing—original draft preparation, A.M.; writing—review and editing, A.M., F.L.M., V.B., F.R. and M.D.V.; visualization, A.M. and F.L.M.; supervision, F.R. and M.D.V.; All authors have read and agreed to the published version of the manuscript.

Funding: This research received no external funding.

Institutional Review Board Statement: Not applicable.

Informed Consent Statement: Not applicable.

Data Availability Statement: Not applicable.

Conflicts of Interest: The authors declare no conflict of interest.

References

1. Cho, N.H.; Shaw, J.E.; Karuranga, S.; Huang, Y.; da Rocha Fernandes, J.D.; Ohlrogge, A.W.; Malanda, B. IDF Diabetes Atlas: Global Estimates of Diabetes Prevalence for 2017 and Projections for 2045. *Diabetes Res. Clin. Pract.* **2018**, *138*, 271–281. [CrossRef]
2. Rajendran, R.; Rayman, G. Point-of-Care Blood Glucose Testing for Diabetes Care in Hospitalized Patients: An Evidence-Based Review. *J. Diabetes Sci. Technol.* **2014**, *8*, 1081–1090. [CrossRef]
3. Kim, J.; Campbell, A.S.; Wang, J. Wearable Non-Invasive Epidermal Glucose Sensors: A Review. *Talanta* **2018**, *177*, 163–170. [CrossRef]
4. Ray, T.R.; Choi, J.; Bandodkar, A.J.; Krishnan, S.; Gutruf, P.; Tian, L.; Ghaffari, R.; Rogers, J.A. Bio-Integrated Wearable Systems: A Comprehensive Review. *Chem. Rev.* **2019**, *119*, 5461–5533. [CrossRef]
5. Kim, J.; Campbell, A.S.; Esteban-Fernández Ávila, B.; Wang, J.; de Ávila, B.E.F.; Wang, J. Wearable Biosensors for Healthcare Monitoring. *Nat. Biotechnol.* **2019**, *37*, 389–406. [CrossRef] [PubMed]
6. Legner, C.; Kalwa, U.; Patel, V.; Chesmore, A.; Pandey, S. Sweat Sensing in the Smart Wearables Era: Towards Integrative, Multifunctional and Body-Compliant Perspiration Analysis. *Sens. Actuators A Phys.* **2019**, *296*, 200–221. [CrossRef]
7. Chung, M.; Fortunato, G.; Radacsi, N. Wearable Flexible Sweat Sensors for Healthcare Monitoring: A Review. *J. R. Soc. Interface* **2019**, *16*. [CrossRef]
8. Brothers, M.C.; DeBrosse, M.; Grigsby, C.C.; Naik, R.R.; Hussain, S.M.; Heikenfeld, J.; Kim, S.S. Achievements and Challenges for Real-Time Sensing of Analytes in Sweat within Wearable Platforms. *Acc. Chem. Res.* **2019**, *52*, 297–306. [CrossRef] [PubMed]
9. Bruen, D.; Delaney, C.; Florea, L.; Diamond, D. Glucose Sensing for Diabetes Monitoring: Recent Developments. *Sensors* **2017**, *17*, 1866. [CrossRef]
10. Juska, V.B.; Pemble, M.E. A Critical Review of Electrochemical Glucose Sensing: Evolution of Biosensor Platforms Based on Advanced Nanosystems. *Sensors* **2020**, *20*, 6013. [CrossRef] [PubMed]
11. Arakawa, T.; Kuroki, Y.; Nitta, H.; Chouhan, P.; Toma, K.; Sawada, S.; Takeuchi, S.; Sekita, T.; Akiyoshi, K.; Minakuchi, S.; et al. Mouthguard Biosensor with Telemetry System for Monitoring of Saliva Glucose: A Novel Cavitas Sensor. *Biosens. Bioelectron.* **2016**, *84*, 106–111. [CrossRef]
12. Heikenfeld, J.; Jajack, A.; Feldman, B.; Granger, S.W.; Gaitonde, S.; Begtrup, G.; Katchman, B.A. Accessing Analytes in Biofluids for Peripheral Biochemical Monitoring. *Nat. Biotechnol.* **2019**, *37*, 407–419. [CrossRef] [PubMed]
13. Gross, T.M.; Bode, B.W.; Einhorn, D.; Kayne, D.M.; Reed, J.H.; White, N.H.; Mastrototaro, J.J. Performance Evaluation of the MiniMed®Continuous Glucose Monitoring System during Patient Home Use. *Diabetes Technol. Ther.* **2000**, *2*, 49–56. [CrossRef] [PubMed]
14. McGarraugh, G. The Chemistry of Commercial Continuous Glucose Monitors. *Diabetes Technol. Ther.* **2009**, *11* (SUPPL.1), S17–S24. [CrossRef]
15. Christiansen, M.; Bailey, T.; Watkins, E.; Liljenquist, D.; Price, D.; Nakamura, K.; Boock, R.; Peyser, T. A New-Generation Continuous Glucose Monitoring System: Improved Accuracy and Reliability Compared with a Previous-Generation System. *Diabetes Technol. Ther.* **2013**, *15*. [CrossRef] [PubMed]
16. Seget, S.; Rusak, E.; Partyka, M.; Samulska, E.; Pyziak-Skupień, A.; Kamińska, H.; Skała-Zamorowska, E.; Jarosz-Chobot, P. Bacterial Strains Colonizing the Sensor Electrodes of a Continuous Glucose Monitoring System in Children with Diabetes. *Acta Diabetol.* **2021**, *58*, 191–195. [CrossRef]
17. Bailey, T.; Bode, B.W.; Christiansen, M.P.; Klaff, L.J.; Alva, S. The Performance and Usability of a Factory-Calibrated Flash Glucose Monitoring System. *Diabetes Technol. Ther.* **2015**, *17*, 787–794. [CrossRef]
18. Jina, A.; Tierney, M.J.; Tamada, J.A.; McGill, S.; Desai, S.; Chua, B.; Chang, A.; Christiansen, M. Design, Development, and Evaluation of a Novel Microneedle Array-Based Continuous Glucose Monitor. *J. Diabetes Sci. Technol.* **2014**, *8*, 483–487. [CrossRef]
19. Bandodkar, A.J.; Jia, W.; Yardimci, C.; Wang, X.; Ramirez, J.; Wang, J. Tattoo-Based Noninvasive Glucose Monitoring: A Proof-of-Concept Study. *Anal. Chem.* **2015**, *87*, 394–398. [CrossRef]
20. Tierney, M.J.; Kim, H.L.; Burns, M.D.; Tamada, J.A.; Potts, R.O. Electroanalysis of Glucose in Transcutaneously Extracted Samples. *Electroanalysis* **2000**, *12*. [CrossRef]
21. Zhao, J.; Lin, Y.; Wu, J.; Nyein, H.Y.Y.; Bariya, M.; Tai, L.C.; Chao, M.; Ji, W.; Zhang, G.; Fan, Z.; et al. A Fully Integrated and Self-Powered Smartwatch for Continuous Sweat Glucose Monitoring. *ACS Sens.* **2019**, *4*, 1925–1933. [CrossRef]
22. Gao, W.; Emaminejad, S.; Nyein, H.Y.Y.; Challa, S.; Chen, K.; Peck, A.; Fahad, H.M.; Ota, H.; Shiraki, H.; Kiriya, D.; et al. Fully Integrated Wearable Sensor Arrays for Multiplexed in Situ Perspiration Analysis. *Nature* **2016**, *529*, 509–514. [CrossRef] [PubMed]

23. Martín, A.; Kim, J.; Kurniawan, J.F.; Sempionatto, J.R.; Moreto, J.R.; Tang, G.; Campbell, A.S.; Shin, A.; Lee, M.Y.; Liu, X.; et al. Epidermal Microfluidic Electrochemical Detection System: Enhanced Sweat Sampling and Metabolite Detection. *ACS Sensors* **2017**, *2*, 1860–1868. [CrossRef]
24. Simmers, P.; Li, S.K.; Kasting, G.; Heikenfeld, J. Prolonged and Localized Sweat Stimulation by Iontophoretic Delivery of the Slowly-Metabolized Cholinergic Agent Carbachol. *J. Dermatol. Sci.* **2018**, *89*, 40–51. [CrossRef] [PubMed]
25. Emaminejad, S.; Gao, W.; Wu, E.; Davies, Z.A.; Nyein, H.Y.Y.; Challa, S.; Ryan, S.P.; Fahad, H.M.; Chen, K.; Shahpar, Z.; et al. Autonomous Sweat Extraction and Analysis Applied to Cystic Fibrosis and Glucose Monitoring Using a Fully Integrated Wearable Platform. *Proc. Natl. Acad. Sci. USA* **2017**, *114*, 4625–4630. [CrossRef]
26. Bhide, A.; Cheeran, S.; Muthukumar, S.; Prasad, S. Enzymatic Low Volume Passive Sweat Based Assays for Multi-Biomarker Detection. *Biosensors* **2019**, *9*, 13. [CrossRef] [PubMed]
27. Yu, H.; Sun, J. Sweat Detection Theory and Fluid Driven Methods: A Review. *Nanotechnol. Precis. Eng.* **2020**, *3*, 126–140. [CrossRef]
28. Perrault, C.M.; Salmon, H.; Mercier, O.; Roy, E. Insights on Polymers for Microfluidics Applied to Biomedical Applications. *Res. Rev. Polym.* **2017**, *8*, 1–7.
29. Young, E.W.K.; Berthier, E.; Guckenberger, D.J.; Sackmann, E.; Lamers, C.; Meyvantsson, I.; Huttenlocher, A.; Beebe, D.J. Rapid Prototyping of Arrayed Microfluidic Systems in Polystyrene for Cell-Based Assays. *Anal. Chem.* **2011**, *83*, 1408–1417. [CrossRef]
30. Nguyen, H.-T.; Thach, H.; Roy, E.; Huynh, K.; Perrault, C.M.-T. Low-Cost, Accessible Fabrication Methods for Microfluidics Research in Low-Resource Settings. *Micromachines* **2018**, *9*, 461. [CrossRef]
31. Clark, L.C.; Lyons, C. Electrode Systems for Continuous Monitoring in Cardiovascular Surgery. *Ann. N. Y. Acad. Sci.* **1962**, *102*, 29–45. [CrossRef] [PubMed]
32. Putzbach, W.; Ronkainen, N.J. Immobilization Techniques in the Fabrication of Nanomaterial-Based Electrochemical Biosensors: A Review. *Sensors* **2013**, *13*, 4811–4840. [CrossRef]
33. Bandodkar, A.J.; Jeang, W.J.; Ghaffari, R.; Rogers, J.A. Wearable Sensors for Biochemical Sweat Analysis. *Annu. Rev. Anal. Chem.* **2019**, *12*, 1–22. [CrossRef]
34. Yoo, E.H.; Lee, S.Y. Glucose Biosensors: An Overview of Use in Clinical Practice. *Sensors* **2010**, *10*, 4558–4576. [CrossRef]
35. Lović, J.; Stevanović, S.; Nikolić, N.D.; Petrović, S.; Vuković, D.; Prlainović, N.; Mijin, D.; Avramov Ivić, M. Glucose Sensing Using Glucose Oxidase-Glutaraldehyde-Cysteine Modified Gold Electrode. *Int. J. Electrochem. Sci* **2017**, *12*, 5806–5817. [CrossRef]
36. Lee, H.; Choi, T.K.; Lee, Y.B.; Cho, H.R.; Ghaffari, R.; Wang, L.; Choi, H.J.; Chung, T.D.; Lu, N.; Hyeon, T.; et al. A Graphene-Based Electrochemical Device with Thermoresponsive Microneedles for Diabetes Monitoring and Therapy. *Nat. Nanotechnol.* **2016**, *11*, 566–572. [CrossRef] [PubMed]
37. Rajaković, L.V.; Marković, D.D.; Rajaković-Ognjanović, V.N.; Antanasijević, D.Z. Review: The Approaches for Estimation of Limit of Detection for ICP-MS Trace Analysis of Arsenic. *Talanta* **2012**, *102*, 79–87. [CrossRef]
38. Wang, J. Electrochemical Glucose Biosensors. *Chem. Rev.* **2008**, *108*, 814–825. [CrossRef]
39. Fois, M.; Arrigo, P.; Bacciu, A.; Monti, P.; Marceddu, S.; Rocchitta, G.; Serra, P.A. The Presence of Polysaccharides, Glycerol, and Polyethyleneimine in Hydrogel Enhances the Performance of the Glucose Biosensor. *Biosensors* **2019**, *9*, 95. [CrossRef] [PubMed]
40. Lee, H.; Hong, Y.J.; Baik, S.; Hyeon, T.; Kim, D.-H. Enzyme-Based Glucose Sensor: From Invasive to Wearable Device. *Adv. Healthc. Mater.* **2018**, *7*. [CrossRef]
41. Sonner, Z.; Wilder, E.; Heikenfeld, J.C.; Kasting, G.; Beyette, F.; Swaile, D.; Sherman, F.; Joyce, J.; Hagen, J.A.; Kelley-Loughnane, N.; et al. The Microfluidics of the Eccrine Sweat Gland, Including Biomarker Partitioning, Transport, and Biosensing Implications. *Biomicrofluidics* **2015**, *9*. [CrossRef] [PubMed]
42. Mano, N. Engineering Glucose Oxidase for Bioelectrochemical Applications. *Bioelectrochemistry* **2019**, *128*, 218–240. [CrossRef]
43. Nguyen, H.H.; Lee, S.H.; Lee, U.J.; Fermin, C.D.; Kim, M. Immobilized Enzymes in Biosensor Applications. *Materials* **2019**, *12*, 121. [CrossRef] [PubMed]
44. Guinovart, T.; Crespo, G.A.; Rius, F.X.; Andrade, F.J. A Reference Electrode Based on Polyvinyl Butyral (PVB) Polymer for Decentralized Chemical Measurements. *Anal. Chim. Acta* **2014**, *821*, 72–80. [CrossRef] [PubMed]
45. Fallahi, H.; Zhang, J.; Phan, H.-P.; Nguyen, N.-T. Flexible Microfluidics: Fundamentals, Recent Developments, and Applications. *Micromachines* **2019**, *10*, 830. [CrossRef]

micromachines

Review

Light-Emitting Textiles: Device Architectures, Working Principles, and Applications

Marco Cinquino [1,2], Carmela Tania Prontera [2,*], Marco Pugliese [1,2], Roberto Giannuzzi [1,2], Daniela Taurino [1], Giuseppe Gigli [1,2] and Vincenzo Maiorano [2]

[1] Department of Mathematics and Physics "Ennio De Giorgi", University of Salento, via Arnesano, 73100 Lecce, Italy; marco.cinquino@unisalento.it (M.C.); marco.pugliese@unisalento.it (M.P.); roberto.giannuzzi@unisalento.it (R.G.); daniela.taurino@unisalento.it (D.T.); giuseppe.gigli@unisalento.it (G.G.)
[2] Institute of Nanotechnology (CNR-NANOTEC), National Research Council, via Monteroni, 73100 Lecce, Italy; vincenzo.maiorano@nanotec.cnr.it
* Correspondence: tania.prontera@nanotec.cnr.it

Abstract: E-textiles represent an emerging technology aiming toward the development of fabric with augmented functionalities, enabling the integration of displays, sensors, and other electronic components into textiles. Healthcare, protective clothing, fashion, and sports are a few examples application areas of e-textiles. Light-emitting textiles can have different applications: sensing, fashion, visual communication, light therapy, etc. Light emission can be integrated with textiles in different ways: fabricating light-emitting fibers and planar light-emitting textiles or employing side-emitting polymer optical fibers (POFs) coupled with light-emitting diodes (LEDs). Different kinds of technology have been investigated: alternating current electroluminescent devices (ACELs), inorganic and organic LEDs, and light-emitting electrochemical cells (LECs). The different device working principles and architectures are discussed in this review, highlighting the most relevant aspects and the possible approaches for their integration with textiles. Regarding POFs, the methodology to obtain side emissions and the critical aspects for their integration into textiles are discussed in this review. The main applications of light-emitting fabrics are illustrated, demonstrating that LEDs, alone or coupled with POFs, represent the most robust technology. On the other hand, OLEDs (Organic LEDs) are very promising for the future of light-emitting fabrics, but some issues still need to be addressed.

Keywords: light-emitting e-textiles; alternating current electroluminescent devices; light emitting diodes; light electrochemical cells; polymeric optical fibers

Citation: Cinquino, M.; Prontera, C.T.; Pugliese, M.; Giannuzzi, R.; Taurino, D.; Gigli, G.; Maiorano, V. Light-Emitting Textiles: Device Architectures, Working Principles, and Applications. *Micromachines* **2021**, *12*, 652. https://doi.org/10.3390/mi12060652

Academic Editors: Maria Serena Chiriacò, Francesco Ferrara and Elisabetta Primiceri

Received: 6 May 2021
Accepted: 29 May 2021
Published: 2 June 2021

1. Introduction

In recent years, the growing innovations in the field of wearable technologies have been changing and improving our everyday lives. Typical examples of wearable and portable devices are in the form of smart watches, glasses, wristbands, and belts. Unfortunately, these kinds of devices are composed of rigid and planar materials, so they may be uncomfortable if worn. In this context, the possibility to integrate microelectronic devices with fabrics represents an important innovation toward the development of a more comfortable and versatile technology [1]. In particular, fabric represents a ubiquitous element of our life and therefore it is the ideal platform for portable and wearable device integration. The idea is to combine the typical textile characteristics such as flexibility, lighter weight, wearability, breathability, and low cost with electronic functionalities. Different sensors have been already integrated into textiles such as pressure sensors [2], strain sensors [3], human stress sensors based on electrochemical transistors [4], glucose electrochemical sensors [5], etc. Power generators and energy storage devices are also fundamental for the advancement of wearable technology. Triboelectric nanogenerators [6], piezoelectric nanogenerators [7], and solar cells based on dyes [8], organic materials [9], and perovskites [10] have been successfully developed. Energy storage devices such as batteries [11] and supercapacitors [12–15] have also been fabricated on textiles.

Light emission is another functionality that can be integrated with textiles with great potential as a visual communication element useful in sensing applications, fashion design, and light therapy. Flexibility, reasonable lifetimes, washability, large display areas, and low production costs are certainly the fundamental requirements for obtaining light-emitting fabrics that are competitive on the market. All these aspects are discussed in this review, and the issues to solve for the future of light-emitting textiles are highlighted. In particular, light emission can be integrated with textiles in different ways: employing self-emitting devices or using optical fibers coupled with external light-emitting devices, such as LEDs. Self-emitting devices can be integrated directly onto the fiber or can be fabricated and/or attached on top of the fabric surface. Different device architectures have been employed, whose working principles, advantages, disadvantages, challenges related to textile integration, and literature results are summarized in Section 2. In Section 2, the optical fibers approach is also described. In Sections 3 and 4, the possible health and environmental concerns related with such technologies and the main application areas are discussed.

2. Light Emitting Technologies: Working Principles and Textile Integration

2.1. ACEL Devices

The phenomenon of alternating current electroluminescence (ACEL) was reported for the first time by Destriau in 1935 [16]. He observed that copper-doped zinc sulfide crystals, suspended in castor oil and sandwiched between two mica sheets, emitted light with the application of a strong alternating electric field. Only in 1952, after the development of transparent conducting electrodes based on tin oxide (SnO_2), the first practical device was prepared [17]. It consisted of a doped ZnS powder embedded in a dielectric matrix and sandwiched between two electrodes. Today, ACEL devices are fabricated in two different ways: in the simplest approach, the dielectric and the phosphor are mixed together and then deposited as a single layer; alternatively, the phosphor can be deposited as thin film and sandwiched between two dielectrics, as reported in Figure 1a [18]. In both cases, from the electrical point of view, the structure of the device is that of a capacitor. The working principle of this kind of device involves different processes: electron injection, electron transport, electron impact and excitation, and finally radiative recombination (Figure 1a) [19]. Specifically, when a high AC voltage is applied, a high electric field is generated between the electrodes. If the electric field is sufficiently high, the electrons at the dielectric-phosphor interface can be injected into the emitting layer (hot injection). The electrons are then accelerated by the electric field toward the opposite electrode, hitting and exciting the phosphor particles. The following radiative relaxation process gives rise to visible light emission. Different phosphor materials can be used and different colors can be obtained by changing the doping impurities. Such devices are usually fabricated by using solution-based deposition techniques such as screen printing in ambient conditions, showing great potential for large-area displays. Considering the working principle, different typologies of electrodes can be used with at least one of the two being transparent without stringent requisite in terms of work function. The total thickness of the device is in the order of tens of microns and such a characteristic confers sufficient flexibility of the device but also good mechanical stability. Regarding the optoelectronic properties, ACEL devices show a wide viewing angle, fast response time, low power consumption, and high contrast in addition to simple manufacturing and cost-effectiveness [18]. On the other hand, such devices show limited luminance values and the high AC voltage values usually employed (50–200 V) can be too hazardous for wearable applications. Furthermore, the ACEL device can be damaged by humidity and for such reason an encapsulation process is necessary [20].

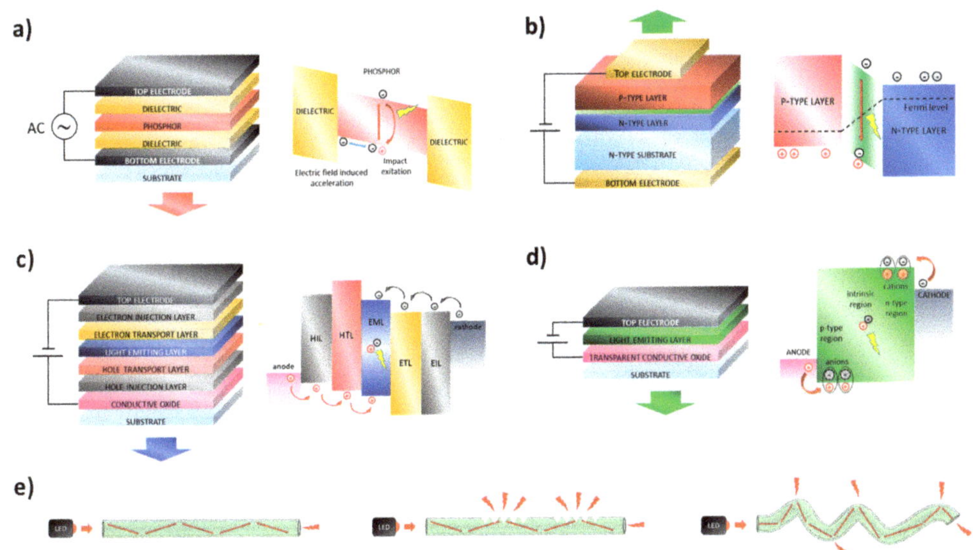

Figure 1. Device architecture and working principle of (**a**) ACEL device; (**b**) LED device; (**c**) OLED device; (**d**) LEC device; (**e**) Schematic representation of distal-end-emitting optical fibers and side-emitting optical fibers by using the microperforation and the macrobending approaches.

Examples of ACEL devices integrated on both single fibers and fabrics are discussed below and they are summarized in Tables 1 and 2. One of the first examples of ACEL devices on single fibers that can be integrated with woven and knitted fabrics was demonstrated in 2012 [21]. The device consisted of a conductive yarn as the electrode, an insulating layer, an EL phosphor layer, and a second electrode based on a conductive yarn helically wound. An automatic dispensing and curing process was employed for the deposition of the insulating and emitting layer. The electro-optical characteristics of the light-emitting fiber are poor and need to be optimized, but the authors reported a high throughput fabrication process of ACEL devices on fiber. Another example of a fiber-based ACEL device created by using a scalable deposition technique was reported by researchers of Drexel University [22–24]. It consists of a slot-die coating system customized for a single fiber. The resulting devices were tested under different mechanical conditions and the authors demonstrated an improvement in strength and robustness when the devices were integrated into a knitted structure [24]. ACEL devices on flexible polyethylene terephthalate (PET) single fibers were also fabricated by dip-coating techniques and an AgNWs solution was employed for the transparent electrode deposition [25]. The device shows a luminance of about 200 cd/m^2 at 195 V and 2 kHz with excellent flexibility, good mechanical characteristics, and wearability. The substitution of the material employed for the fiber fabrication can give an added value to the device. Flexible and water dissolvable PVA (poly(vinyl alcohol)) fibers were used for ACEL devices that can completely disintegrate after 180 min of water immersion with great potential in the reduction of e-waste [26]. A stretchable ACEL device (Figure 2a) was demonstrated by using a PDMS polydimethylsiloxane (PDMS) fiber with luminance values that do not change with stretching strain [27]. Stretchable single-fiber ACEL devices were also fabricated by the group of Peng by using two different methods [28,29]. The first one consists of 3D printing an elastomeric light-emitting tube composed of ZnS phosphor powders and silicon elastomer sandwiched between two aligned carbon nanotube sheets (Figure 2e) [28]. Such a device can be stretched up to 200% and shows a luminance of about 15 cd/m^2 at 6.4 V/μm and 1500 Hz. The second method is based on the fabrication of a stretchable electroluminescent fiber via a continuous one-step extruding process. The fiber is composed of two inner hydrogel electrodes and a ZnS/silicon elastomer blend; it shows a luminance of about 250 cd/m^2 at 8 V/μm and 1500 Hz when the 40% of ZnS is employed and it could be stretched up to 800% [29]. Graphene is another possible transparent electrode that can be interesting for light-emitting devices and the group of Craciun developed an ACEL device on top of tape-shaped graphene-covered fiber, the architecture of which is shown in Figure 2d [30].

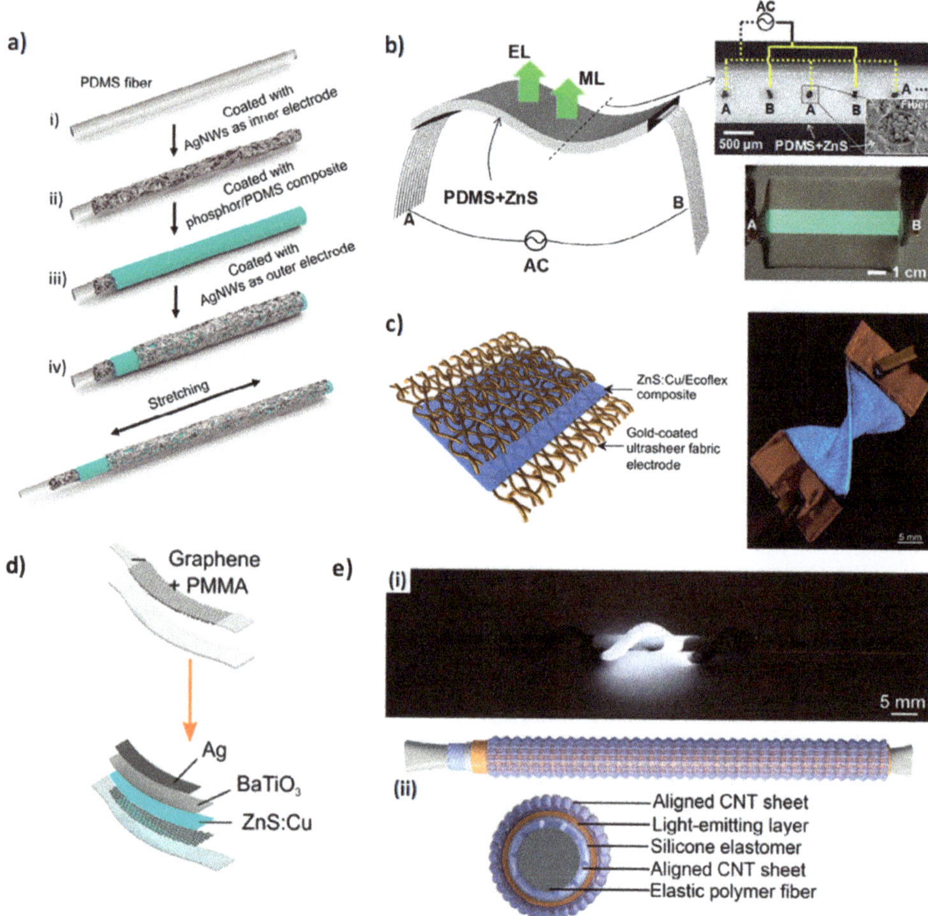

Figure 2. (**a**) Scheme of the fabrication process of a stretchable ACEL fiber. (**b**) Picture of a textile-fiber-embedded ACEL device, optical and scanning electron microscope cross-sectional images, and picture of the ACEL device under 500 V at 1 KHz. (**c**) Device structure based on ZnS:Cu/Ecoflex composite sandwiched between two gold-coated ultrasheer fabric electrodes and photograph of the light-emitting textile. (**d**) Graphene-based ACEL device. (**e**) Picture of a fiber-based ACEL device wrapped around a glass bar and scheme of the device structure. (**a**) Reproduced under the terms of the CC BY 4.0 license [27]. (**b**) Reproduced with permission [31]. Copyright 2020, Elsevier. (**c**) Reproduced with permission [32]. Copyright 2020, Elsevier. (**d**) Reproduced under the terms of the CC BY 4.0 license [30]. (**e**) Reproduced with permission [28]. Copyright 2018, Royal Society of Chemistry.

Table 1. Deposition techniques, structure, and performances of ACEL devices on a single fiber.

Year	Deposition Techniques	Device Structure	Performances	Ref.
2012	Automatic dispensing and curing process	Silver-coated yarn/DuPont dielectric paste/DuPont phosphor ink/silver-coated yarn	1.3 cd/m^2 at 370 V and 2 kHz	[21]
2017	Slot-die coating	Silver-coated yarn/DuPont dielectric paste/DuPont phosphor ink/silver-coated yarn	50 cd/m^2	[22]
2017	Slot-die coating	Silver-coated yarn/DuPont dielectric paste/DuPont phosphor ink/DuPont conductive paste	n.a.	[23]
2017	Dip coating	PET fiber/AgNWs/silicone/phosphor/AgNWs/silicone	202 cd/m^2 at 195 V and 2 kHz	[25]
2018	Slot-die coating	Silver-coated yarn/DuPont dielectric paste/DuPont phosphor ink/DuPont conductive paste	50 cd/m^2 at 100 V and 20 kHz	[24]
2018	Dip coating	PDMS fiber/AgNWs/phosphor:PDMS/AgNWs	100 cd/m^2 at 400 V and 1 kHz	[27]
2018	3D printing and automatic wrapping	Elastic polymer fiber/Aligned CNT sheet/Silicone elastomer/Light-emitting layer/Aligned CNT sheet	15 cd/m^2 at 6.4 V/μm and 1.5 kHz	[28]
2018	Extrusion	Two inner hydrogel electrodes + ZnS:silicone elastomer	250 cd/m^2 at 8 V/μm and 1.5 Hz	[29]
2018	Spin coating	Graphene/ZnS:Cu/BaTiO3/Ag	n.a.	[30]
2019	Dip coating	PVA fiber/AgNWs/ZnS:Cu+PVP/AgNWs	100 cd/m^2 at 300 V and 0.4 kHz	[26]

As already mentioned in the introduction, the light-emitting device can also be fabricated directly on top of the fabric. ACEL devices were fabricated on fabric by using different deposition techniques and different transparent electrodes. Screen printing is the most used deposition technique for ACEL fabrication and it consists of transferring an ink onto a substrate through a mesh, changing the mesh to deposit different patterns. It was demonstrated that dispensing printing can be employed as an alternative deposition technique with comparable performance and with a versatility improvement since the deposition pattern can be changed digitally with a potential reduction of time and costs [33]. ACEL devices were also fabricated on semitransparent textiles (38% light transmission) by using a bottom emission configuration [34]. Inkjet printing and dispensing printing were employed to deposit poly(3,4-ethylenedioxythiophene (PEDOT:PSS) as a bottom transparent electrode and all the other layers, respectively. The device shows a maximum luminance of 44 cd/m^2 at 400 V and 400 Hz. A bilayer of carbon nanotubes (CNTs) and PEDOT:PSS was also used as a bottom electrode for the fabrication of an ACEL device on top of a transparent PET mesh fabric [35]. CNTs were deposited via dip coating, while PEDOT:PSS was inkjet-printed and finally a nozzle extrusion system was employed for the deposition of dielectric and phosphor layers. The electroluminescent device reached a luminance of 75 cd/m^2 at 200 V and 3.5 KHz. Other carbon nanomaterials were also employed as transparent electrodes [36,37]. For the deposition of the carbon nanomaterials (carbon nanotubes and graphene platelets), poly(methyl methacrylate) (PMMA) was first employed as polymeric carrier with poor results on the mechanical stability of the device. [36] More recently, a best result was reached by using a thermoplastic polyurethane polymer as an alternative carrier and the corresponding device had a luminance value of about 200 cd/m^2 at 160 V and 1000 Hz and it did not exhibit visible degradation after 20 washing cycles [37]. An ACEL device was also successfully realized on top of a stretchable textile by using polypyrrole as a transparent conductive layer and a silicone elastomer as a matrix for the phosphor powder [38]. The resulting device had a luminance value of 350 cd/m^2 that was preserved after hundreds of stretching cycles. An innovative approach for the fabrication of fabric-based ACEL devices was reported by the group of Carmichael [32,39]. They obtained a conductive stretchable textile by using a solution-based metallization process of knitted textile, demonstrating that the coating was conformal and the textile voids remained intact, thus preserving the stretchability [39]. The conductive knitted fabric showed a resistance reduction of 80% when the fabric was stretched to 15% and it maintained such resistance at 160% of strain; furthermore, its resistance was preserved after 10 washing cycles. The conductive substrate was tested as an electrode for ACEL devices by depositing on top of it $BaTiO_3$+PDMS, ZnS:Cu+PDMS and PEDOT:PSS; a uniform blue emission was visible at 165 V and 37 kHz that was stable below 40% of strain [39]. An upgrade of this device architecture was recently reported and it consisted of a ZnS:Cu/Ecoflex composite sandwiched between two gold-coated fabric electrodes, as shown in Figure 2c. The device showed a uniform blue emission even after the application of a strain of 200% [32]. Another methodology for ACEL integration on textile was developed in 2019 and it can be considered a hybrid fiber/textile approach with improved mechanical stability [31]. The device consisted of Ag-coated nylon fibers embedded into a composite based on PDMS and ZnS phosphors (Figure 2b); when an AC voltage is applied the light is emitted from the ZnS particles that surround the fibers. The resulting coplanar structure showed high durability and stable electrical conductivity.

From the results reported in this paragraph, it is evident that ACEL technology is robust enough for textile integration and high scalable deposition techniques can be employed for fabrication with the possibility of low-cost production. Although researchers are currently working to improve the potentiality of ACEL technology in the wearable electronic field [40], at the moment low luminance and high operating voltage represent the main factors limiting the development of this technology.

Table 2. Deposition techniques, structure, and performances of ACEL devices on fabric.

Year	Deposition Techniques	Device Structure	Performances	Ref.
2011	Inkjet printing	PEDOT:PSS/Phosphor:epoxy resin/aluminium	44 cd/m^2 at 400 V and 0.4 kHz	[34]
2012	Inkjet printing + nozzle extrusion	CNT and/or PEDOT:PSS/dielectric layer/phosphor layer/aluminium	70 cd/m^2 at 200 V and 3.5 kHz	[35]
2014	Screen printing	Silver/dielectric layer/luminophore/CNT-GNP electrode	n.a.	[36]
2016	Dispenser printing	FabInks bottom electrode/FabInks dielectric/FabInks phoshor/ FabInks transparent tonductor	300 cd/m^2 at 370 V and 1 kHz	[33]
2017	Chemical deposition + casting	Polypirrole/ ZnS:Silicone elastomer/hydrogel film	350 cd/m^2 at 5 V/µm and 2 kHz	[38]
2018	Solution-based metallization + spin coating	Gold-coated textile/BaTiO3+PDMS/ZnS:Cu+PDMS/PEDOT:PSS	n.a.	[39]
2019	Embedding of fibers into phosphor:PDMS composite	Ag-coated fibers/ZnS phosphor+PDMS	35 cd/m^2 at 1.8 V/µm and 2 kHz	[31]
2019	Screen printing	Graphene-based electrode/BaTiO3/ZnS:Cu/BaTiO3/CNT or ATO transparent electrode	300 cd/m^2 at 160 V and 2 kHz	[37]
2020	Solution-based metallization + spin coating + lamination	Gold-coated textile/ZnS:Cu+Ecoflex/gold-coated textile	n.a.	[32]

2.2. LED Devices

Unlike ACEL devices, LEDs (light-emitting diodes) need a DC voltage of only a few volts to emit light. The first practical inorganic LED in the visible region was reported in early 1962 and it was based on $Ga(As_{1-x}P_x)$ p-n junction [41]. A diode is a two-terminal electronic component that can conduct current only in one direction and, if specific light-emitting materials are employed, the current flux induces the formation of excitons that radiatively decay. Specifically, an LED is a p-n junction: when a voltage is applied, electrons and holes are pushed through the n and p regions, respectively, and they reach the "active region" that is close to the junction [42]. Electrons move into the conduction band, while holes move into the valence band; when electrons and holes recombine in the active region a radiative decay can occur, whose wavelength emission is related to the semiconductor band gap. A scheme of the working principle is shown in Figure 1b. Compounds based on Ga, Al, In, N, P, and As (III-V materials) are the core of inorganic LEDs [43]; such materials are crystalline in form and are usually deposited by using high temperature and high vacuum on sapphire substrates, which induce epitaxial growth. Since small defects can strongly affect the device characteristics and due to the high costs of the fabrication processes, the production of small-sized devices is usually more convenient. Nowadays, LEDs represent commercially available and high-performance light-emitting devices with dimensions of a few millimeters. However, LEDs do not exhibit the flexibility and stretchability properties desirable for textile integration. Nonetheless, this technology is reliable and numerous research efforts were made to integrate such devices with textiles. In particular, two different approaches are reported in the literature; the first approach involves the fabrication of a textile electronic circuit and the subsequent attachment of different electronic devices onto the circuit, including LEDs. Alternatively, the device can be directly embedded into the fiber; by using this approach, the device results are less visible to the viewer, the fabric is more comfortable, and the electrical interconnections are more mechanically robust. Regarding the first approach, two aspects need to be considered: the fabrication of the textile electrical circuits and their interconnection with external devices that can also be rigid. Different methods for the fabrication of electrical circuit have been reported in the literature: embroidering of conductive yarns [44,45], weaving [46,47] and knitting by using conductive and non-conductive yarns [48], laser cutting of conductive fabrics [49], screen printing [50], and use of plastic-based e-strips that can be woven like traditional yarns [51,52]. About the interconnection between the circuits and the external devices: soldering [49,51,52], gluing with conductive [46,52] and non-conductive adhesives [47], embroidering and stitching with conductive yarns [45,49], and crimping [53] are the most used techniques. In 2006, Buechley created an "e-textile construction kit" containing a microcontroller, sensors, and actuators (including LEDs), an infrared transceiver, an on/off switch, and a battery pack [44]. All the components were made of fabric or were designed to be easily stitched onto fabric; furthermore, the different elements can be connected through sewn conductive traces. The fabric circuit was obtained by using a methodology that combines the "iron-on" technique and laser cutting; firstly, a heat-activated paper adhesive is attached to a conductive fabric, afterward, a laser cutter is used to etch the circuit pattern into the fabric and the paper is removed from underneath the circuit, finally, the circuit is ironed onto the second piece of fabric [44,49]. The highly scalable and low-cost screen-printing technique was exploited by Kim et al. as a valid alternative to laser cutting, and the flip-chip method was employed to interconnect external electrical elements to the circuit [50]. The connection to the textile circuit can be also carried out by using a thermoplastic non-conductive adhesive [47,54]; by heat and pressure application the adhesive melts and electrical contact is formed. Stable mechanical and electrical contact is obtained after cooling. This method can be used with various types of textiles, and it represents a very easy and reliable approach. Package dies and more specifically LEDs can be also integrated via weaving. In particular, Parvoka et al. reported a specific weaving method able to integrate and hide the LED into the textile and this aspect is interesting to improve the aesthetics of the product (Figure 3a) [55,56].

Plastic-based e-fibers that can be woven into textiles by using a commercial manufacturing process were also reported [51]. The e-fibers were fabricated on top of a flexible plastic substrate by using standard microfabrication techniques and, afterward, the substrate was cut into 5-cm-long and <2-mm-wide strips. The obtained e-fibers were exposed to bending radii of 160 micron and tensile strain of about 20% without failure. In order to demonstrate the robustness of this approach, temperature and humidity sensors as well as LED devices were integrated on the e-fibers by soldering. A similar approach was recently used for the fabrication of e-strips on top of flexible plastic substrates, but in this case the weaving process was designed to conceal its presence from the wearer [52]. In particular, during the weaving process, bespoke pockets were obtained for the placement of the circuit; this methodology reduces the visibility of the circuit and makes it more comfortable when in contact with the skin. Also, in this case, external devices such as LEDs and microcontrollers were attached on top of the circuit by using a solder paste or anisotropic conductive paste. Moreover, the filament circuits embedded within the textile show better mechanical and durability performances compared to the corresponding filament that is not integrated into the fabric.

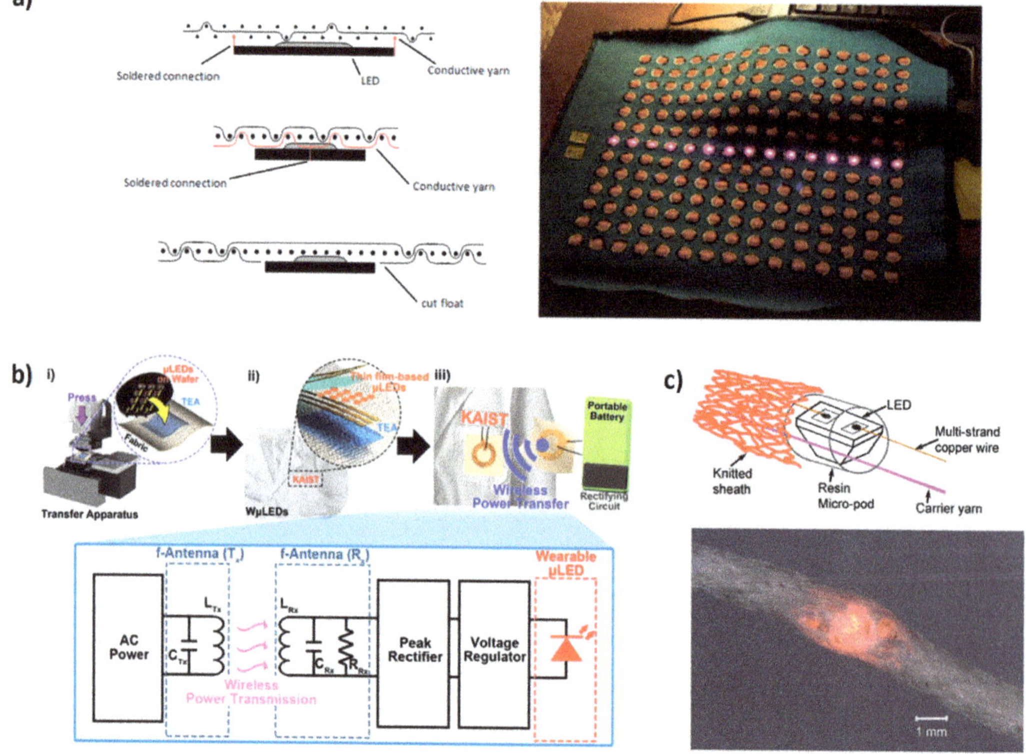

Figure 3. (**a**) Schematic illustration of LED integration with textile and picture of a textile LED display. (**b**) Representation of wireless-powered wearable μLEDs with a scheme of the circuit. (**c**) Scheme of the E-yarns structure with an LED and related picture. (**a**) Reproduced with permission [56]. Copyright 2013, NC State University. (**b**) Reproduced with permission [57]. Copyright 2018, Elsevier. (**c**) Reproduced under the terms of the CC BY 4.0 license [58].

MicroLEDs are an alternative to traditional LEDs for textile integration. They show high power efficiency, excellent stability, and their small dimensions do not compromise the flexibility and conformability of a typical textile. Lee et al. demonstrated a wireless-powered wearable microLED transferred on 100% cotton textile by using a transparent elastomeric adhesive, as shown in Figure 3b [57]. The device was resistant to stretching and bending, and it was also stable under high temperature and high humidity conditions (85 °C/85% RH).

As already mentioned, a different approach can be pursued in which the device is directly incorporated into the fiber. A thermal drawing process was reported in 2018 by Rein and coworkers [59]. This method is highly scalable and it allows for the creation of electrically connected diode fibers that maintain excellent performance after ten washing cycles. Another highly scalable method was reported by Hardy and coworkers, and it consists of a multistep process: soldering of dies onto a copper wire, encapsulation of the soldered die in a polymer micropod and finally twisting of textile yarns around the copper wire and the encapsulated dies (Figure 3c) [58,60,61]. Such a method was used for the embedding of both LEDs and photodiodes and it allows for a high-speed fabrication of e-yarns, which can be easily integrated into textiles by using the traditional equipment of the textile industry.

Considering all the reported literature results, it is possible to conclude that LEDs are one of the best performing light-emitting technologies that can be integrated with textiles as a highly luminous, rigid, and point-like light source. Unfortunately, LEDs are not suitable for the fabrication of large-area and flexible light-emitting devices on textiles, and other technologies should be evaluated.

2.3. OLED Devices

OLEDs (Organic Light Emitting Diodes) are an evolution of LED devices in which the semiconductors are organic materials. The first OLED was fabricated by Tang and Van Slyke in 1987 and it was composed of two organic layers, an aromatic diamine as the hole transport layer and 8-hydroxyquinoline aluminum as an emissive layer, confined between indium tin oxide (ITO) and a Mg:Ag electrode [62]. In 1990, Burroughes et al. reported a high-efficiency green light-emitting polymer-based OLED by using poly(p-phenylene vinylene) as an emissive material [63]. To improve light emission efficiency, phosphorescent and thermally activated delayed fluorescent (TADF) materials have been developed and studied and nowadays they are largely applied as active materials in OLEDs [64–66]. The most employed deposition technique for OLED manufacturing is vacuum thermal evaporation, but organic materials can be easily processed by solution-based depositions, and for such reasons they can in principle be fabricated on a large area by using low-cost techniques, differently from inorganic LEDs. The deposition processes and the post-treatments of organic materials do not require high temperatures and it demonstrates the compatibility of such fabrication processes with flexible plastic substrates. Moreover, organic materials are intrinsically flexible, and this represents an advantage for flexible and textile-based device manufacturing in terms of mechanical and electrical stability. On the other hand, electrodes typically used in OLEDs, such as metals and ITO, can be damaged by mechanical deformations. Indeed, more flexible electrode materials are being studied [67]. Another important aspect of the fabrication of flexible and textile based OLEDs is their thickness. OLEDs show a total thickness in the order of 100–200 nm, much thinner compared to ACEL devices and so each step of the fabrication process should be opportunely evaluated to avoid short circuits. In particular, the flatness of the substrate on which the device is deposited is a fundamental requirement. OLEDs for textile integration can be fabricated on single fibers or on top of textiles as planar devices. In the case of the integration of the light-emitting diode on a single fiber, it is sufficient to select a flexible and robust fiber with a smooth surface. On the other hand, planarization is mandatory to correctly fabricate a planar OLED on fabric because textiles show a waviness in the order of microns. Another important aspect is the high sensitivity of OLEDs to oxygen and water vapors that strongly reduce their lifetime. For such reasons, effective encapsulation is a crucial element.

The most important results about the integration of OLED devices with textiles will be illustrated below, both for single-fiber devices and planar OLEDs (see Tables 3 and 4). The first OLED on a single fiber was fabricated by O'Connor et al. in 2007 [68]. Specifically, all layers were deposited via thermal evaporation on a polyimide-coated silica fiber, producing devices with an external quantum efficiency (ηEQE) between 0.07% and 0.15%. Despite the low efficiencies and although the device was made on a silica fiber and not on a textile one, the importance of this work lies in the fact that it represents the first successful attempt to deposit an OLED on a fiber substrate. Another OLED fabricated all-around a fiber was developed by Kwon et al. in 2015 [69]. Here, an anode and active layer were deposited through dip coating to coat the fiber concentrically while, for cathode deposition, LiF/Al were thermally evaporated on the emissive layer. With this new configuration, a luminance of 1458.8 cd/m^2 and a current efficiency (CE) of 3 cd/A were achieved. In 2018, the same group reported another OLED device on a single fiber with an inverted structure [70]. The cathode, the electron injection layer (EIL), and the active material were deposited through dip coating while the hole injection layer (HIL) and the anode were thermally evaporated. A schematization of the fabrication process is shown in Figure 4a. This new configuration exhibited a luminance of 11,780 cd/m^2 and CE of 11.1 cd/A. Additionally, after encapsulation with a 50-nm-thick Al_2O_3 layer, fiber devices showed operating lifetimes of approximately 80 h comparable to the control devices. During the same year, another OLED device on fiber was reported by Ko et al. [71]. The most interesting aspect of this work was the employment of a hollow-core fiber as a substrate. This choice allows for the suppression of the wave-guided light loss at the electrode/substrate interface and avoids the use of additional internal and external light extraction structures (Figure 4d). Thanks to these advantages, the hollow-fiber OLEDs achieved a luminance of 6300 cd/m^2 and CE of 11 cd/A. Moreover, the insertion of an optically active solution into the empty core of the fiber can be exploited to tune the emission wavelength of the device.

An important improvement in fiber OLED efficiencies was achieved by Ko et al. in 2020 [72]. In particular, using a hybrid PEDOT:PSS/Ag fiber as a transparent conducting electrode (TCE) embedded in the polymeric substrate, it was possible to fabricate a fiber OLED characterized by a luminance of 4200 cd/m^2, a CE of 39.6 cd/A, and an EQE of 11.3%, performances comparable to that of the corresponding planar ITO-OLED. In the same year, the fabrication of quantum-dot light-emitting diodes (QLEDs) on a single fiber was reported by Lee et al. [73]. Using CdSe/ZnS and CdS/ZnS core-shell QDs, devices with three different emitting colors (red, green, and blue) were obtained with good performances (340 cd/m^2 at 13 V—blue QLED, 2044 cd/m^2 at 10 V—red QLED, 2240 cd/m^2 at 10 V—green QLED). Furthermore, thanks to the mechanical stability of the fiber, all QLEDs keep 72% lighting at 90° bending, as is visible in Figure 4c. OLED pixels on stripe-shaped PET fibers were also fabricated and were subsequently assembled with conductive fibers to obtain a passive matrix-based textile [74]. The OLED was based on thermal evaporated phosphorescent material as an emitting layer and it showed a maximum current efficiency of about 46 cd/A. Moreover, the woven OLED textile was stable in water and under applied tensile force. The most recent work on fiber OLEDs reports about the deposition of a phosphorescent active layer via dip-coating technique [75]. In particular, three different phosphorescent materials were tested as active layers to prepare phosporescent OLEDs (phOLEDs) with three different emitting colors (red, green, and blue). The resulting devices showed a maximum luminance of 4462, 11,482, and 1199 cd/m^2 and a CE of 16.3, 60.7, and 16.9 cd/A for red, green, and blue phOLEDs, respectively. Moreover, after encapsulation with a 50-nm-thick Al_2O_3 layer, these fiber phOLEDs showed a storage lifetime longer than 4 days and an operating lifetime longer than 10 h.

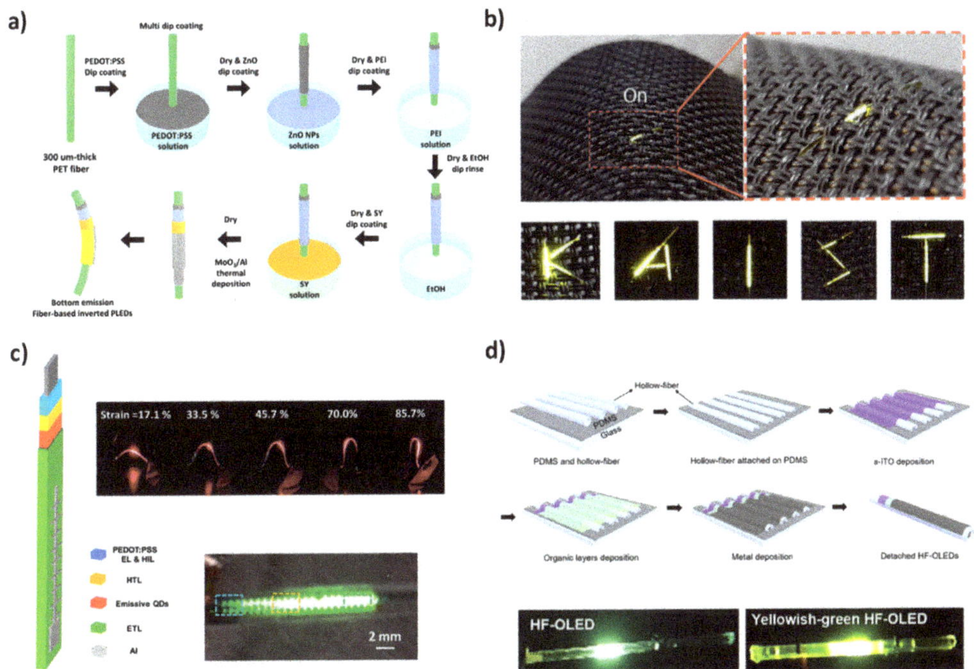

Figure 4. (**a**) Scheme of OLED fabrication on fibers. (**b**) Integration of a fiber-based OLED with a textile. (**c**) Device architecture of the quantum-dot light-emitting diode and pictures of the device in working conditions. (**d**) Fabrication process of the fiber-shaped OLED and device pictures. (**a**) Reproduced with permission [70]. Copyright 2018, American Chemical Society. (**b**) Reproduced with permission [70]. Copyright 2018, American Chemical Society. (**c**) Reproduced with permission [73]. Copyright 2020, American Chemical Society. (**d**) Reproduced with permission [71]. Copyright 2018, Royal Society of Chemistry.

Table 3. Structure and performance of OLEDs on fiber.

Year	Deposition Techniques	Device Structure	Performances	Ref.
2007	Thermal evaporation	Al/Ni/CuPc/NPD/Alq_3/LiF/Al	η_{EQE} 0.07 ÷ 0.15% at 0 ÷ 10 V	[68]
2015	Dip coating + thermal evaporation	PEDOT:PSS/Super Yellow/LiF/Al	1458.8 cd/m^2at 10 V 3 cd/A at 6 V	[69]
2018	Dip coating + thermal evaporation	PEDOT:PSS/ZnO NPs/PEI/Super Yellow/MoO_3/Al	11,780 cd/m^2 at 10 V 11.1 cd/A at 5 V	[70]
2018	Thermal evaporation	ITO/2-TNATA/NPB/Alq_3/LiF/Al	6300 cd/m^2at 13 V 11 cd/A at 12 V	[71]
2020	Spin coating + thermal evaporation	Hybrid fiber TCEs/PEDOT:PSS/PVK:TPD:PBD:Ir(mppy)$_3$/ TPBi/LiF/Al	4200 cd/m^2 at 12 V, 39.6 cd/A and 11.3% of EQE at 7 V	[72]

Table 3. *Cont.*

Year	Deposition Techniques	Device Structure	Performances	Ref.
2020	Dip coating + thermal evaporation	PEDOT:PSS PH1000/PEDOT:PSS AI4083/TFB/QDs/AlZnO/Al	340 cd/m^2 at 13 V for CdS/ZnS (blue QLED) 2044 cd/m^2 (red QLED) and 2240 cd/m^2 (green QLED) at 10 V for CdSe/ZnS	[73]
2020	Thermal evaporation	ITO/HAT-CN/TAPC/TCTA:Ir(ppy)$_2$acac/B3PYMPM/Liq/Al	2900 cd/m2 at 5 V, 46 cd/A at 2.4 V	[74]
2021	Dip coating + thermal evaporation	PEDOT:PSS/ZnO NPs/PEI/PVK:26DCzppy:Ir(ppy)3 (30:30:1 weight ratio)/TCTA/MoO3/Al—green	11,482 cd/m^2 at 6.5 V 60.7 cd/A at 4.5 V for the green OLED	[75]
		PEDOT:PSS/ZnO NPs/PEI/PVK:TPBi:Hex-Ir(phq)2acac (25:25:1 weight ratio)/TCTA/MoO3/Al—red	4462 cd/m^2 at 7 V 16.3 cd/A at 4.5 V for the red OLED	
		PEDOT:PSS/ZnO NPs/PEI/PVK:26DCzppy:Ir(Fppy)3 (30:30:1 weight ratio)/TCTA/MoO3/Al—blue	1199 cd/m^{2}at 6 V 16.9 cd/A at 4 V for the blue OLED	

As for the fabrication of planar OLEDs on textiles, the first attempt was reported by Janietz et al. in 2013 [76]. Specifically, 1 cm^2 orange and green OLEDs were integrated into textiles like spacer warp knitting. The green and orange OLEDs showed a brightness of 1900 and 1100 cd/m^2 and CE of 11.3 and 2.5 cd/A, respectively. Nevertheless, despite the good efficiencies, both OLEDs were not deposited directly on fabric but just integrated after their fabrication. To the best of our knowledge, the first OLED fabricated directly on a fabric substrate was produced by researchers of KAIST in 2013 [77]. An all top-emitting OLED was deposited through thermal evaporation on a plain-woven fabric made of polyester fibers. Before OLED fabrication, the fabric surface was planarized by depositing two polymeric ductile materials, polyurethane (PU) and poly(vinyl alcohol) (PVA), via lamination and spin coating. Thanks to the planarization, a reduction of the roughness from 10 µm to 0.386 µm was reported. OLEDs fabricated on planarized fabric showed a luminance of 7000 cd/m^2 and CE of around 8 cd/A, comparable with performances of the devices fabricated on glass substrates. In 2014, the same group proposed another OLED device fabricated on textiles with an improved planarization process and a multilayer of 3.5 dyads of aluminum oxide (Al_2O_3) and PVA for the encapsulation [78]. They reported a planarization technique employing two layers of PU at low and high viscosity to reduce the large-scale roughness and to ensure a flat surface. The device was obtained via thermal evaporation, and 2 dyads of NPB (N,N′-Di(1-naphthyl)-N,N′-diphenyl-(1,1′-biphenyl)-4,4′-diamine) and WO_3 were deposited as a capping and protective layer on top of the device. Al_2O_3 and PVA were deposited via atomic layer deposition (ALD) and spin coating, respectively. The OLEDs reached a current efficiency of 3 cd/A and luminance of 2000 cd/m^2 with similar performance after 400 h in ambient air, proving the good performance of the encapsulation layer. The same OLED, with the same encapsulation and planarization layer, was also proposed with a new capping/protective dyad composed by NPB and ZnS [79]. ZnS was selected because can be deposited in situ while developing the OLED, it also acts as a moisture barrier, and, thanks to its refractive index (~2.3), allows for a high light extraction in the final OLEDs. The resulting devices reached a maximum luminance of ~1500 cd/m^2 and a maximum current efficiency of ~5 cd/A. Moreover, the barrier properties of the encapsulating multilayer (WVTR = 1.8×10^{-5} g/m^2/day at 30 °C and 90% RH) ensured good performances after 3500 h in ambient air. The same technology was also used to develop a bottom emitting OLED by depositing the anode (PEDOT:PSS) and the active layer (Super Yellow) through spin coating, allowing for the device to partially overcome the disadvantages of thermal evaporation, which is expensive and requires a high vacuum [80]. The result is an OLED with a maximum luminous efficiency and maximum power efficiency of 9.72 cd/A and 7.17 lm/W, respectively, which are approximately 20% lower than those found on the glass reference. An OLED based on a phosphorescent emitting material was also fabricated on planarized fabric (Figure 5a) with a luminance of ~93,000 cd/m^2 and CE of ~49 cd/A [81]. In the study, researchers also reported a new multibarrier encapsulation composed of Al_2O_3 and a soft silane-based polymer that provides an excellent optical transmittance (>90% in the visible region) and an outstanding water vapor transmission rate (WVTR) of 10^{-6} g/m^2/day, nearly identical to that of commercial glass lids. The WVTR of the multilayer barrier remains almost unvaried after 1000 cyclic bends with a radius of 2 cm. Researchers of KAIST also reported a new approach for the fabrication of OLEDs on textiles by using an ultrathin planarization layer and a strain buffer [82]. In particular, they used a new surface-replicating method with a sacrificial layer to obtain a surface with the same flatness of glass. All the phases of the planarization process are shown in Figure 5d. Thanks to this planarization process it is possible to fabricated OLEDs on every kind of textile with performances comparable to that obtained on glass. Additionally, thanks to the Al_2O_3/ZnO encapsulation multilayer that showed a WVTR of 7.87 × 10^{-6} g/m^2/day, the textile OLEDs can retain more than 90% of their luminance after 200 min underwater.

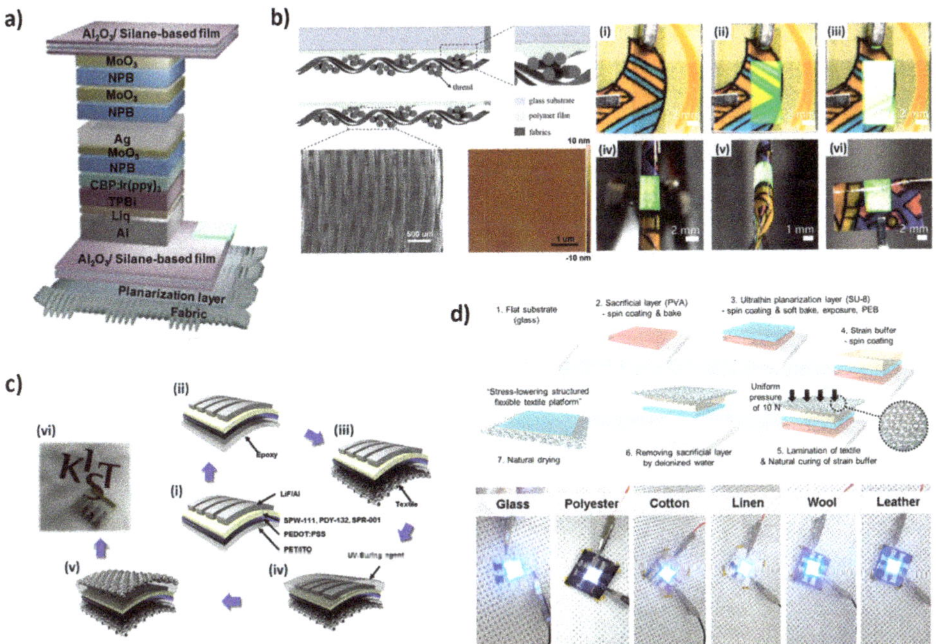

Figure 5. (**a**) Schematic illustration of the fabric-based OLED with planarization and encapsulation multilayer. (**b**) Fabric planarization process by using a polymer film and a glass substrate, SEM picture of the fabric and AFM image of the planarized fabric, photographs of the OLED device on fabric at different driving voltages and under bending conditions. (**c**) Scheme of the fabrication process of a textile/PLED/textile structure. (**d**) Representation of a planarization process based on a replicating method with a sacrificial layer and pictures of OLEDs fabricated on different kinds of planarized textiles. (**a**) Reproduced under the terms of the CC BY 4.0 license [81]. (**b**) Reproduced with permission [83]. Copyright 2020, Elsevier. (**c**) Reproduced with permission [84]. Copyright 2016, Elsevier. (**d**) Reproduced under the terms of the CC BY 4.0 license [82].

A different approach for fabric planarization was reported by the group of Hong-Bo Sun in 2017 [85]. A commercially available photopolymer was deposited via spin coating on top of a silk substrate, providing a good surface morphology (root mean square (RMS) = 0.682 nm) but also high flexibility and mechanical robustness. The device on top of a planarized silk substrate was deposited via thermal evaporation by using an iridium complex as an emitting material and the OLED showed a maximum luminance and current efficiency of 45,545 cd/m^2 and 37.7 cd/A, respectively. Additionally, not only were OLED emission results uniform and defectless, but even after 100 bending cycles no deterioration in luminance and efficiency was observed. The same group improved the performance of the planarization layer by using the same photopolymer previously reported but with a new template-stripping deposition process [86]. They fabricated an OLED on the planarized textile, characterized by a luminance of 15,000 cd/m^2 and CE of around 78 cd/A and outstanding bending stability. The variations of luminance and CE are only 1.7% and 8%, respectively, after 1000 bending cycles at a 1 mm bending radius, which is the best bending stability reported so far. These performances, comparable with those of conventional planar devices, can be ascribed to the flat and smooth planarization layer. They also reported a highly transparent OLED on planarized nylon by using ultrathin metal films as an anode and cathode and by employing a capping layer of NPB [83]. With this new architecture, the device reached a maximum luminance larger than 10,000 cd/m^2 and a CE of 16.7 cd/A. Unfortunately, due to the small thickness of the metal electrodes (7 nm of Au anode and 9 nm of Ag cathode), the OLED has poor bending stability and degrades quickly after few bending cycles.

An innovative and environmentally friendly textile substrate for OLED integration was developed by Park et al., and it was based on a keratin/PVA nanofiber mixture [84]. The fabric exhibited high optical transparency (transmittance of ~85% at λ = 550 nm) and surface roughness with an RMS of 142.7 nm and an arithmetical mean height (Ra) of 104.2 nm. OLEDs fabricated on PET were sandwiched between two sheets of biocompatible fabric (Figure 5c) and in these conditions the device exhibited a luminance of 2781 cd/m^2, 2430 cd/m^2, and 6305 cd/m^2 and maximum current efficiency of 0.29 cd/A, 0.10 cd/A, and 0.38 cd/A for white, red, and yellow emissions, respectively. Although the device was not directly fabricated on top of the textile substrate, this result is interesting for the future development of eco-friendly textile electronic devices.

A well-performing encapsulation process for textile-based organic device fabrication was recently proposed by Jeong et al. [87]. The encapsulation consists of a nano-stratified barrier with an SiO_2 polymer composite, and it was obtained by using ALD and spin-coating techniques. A polymeric solar cell and an OLED were fabricated on the barrier-coated textile, and they preserved almost the same initial performances after several washing tests.

Not only simple OLED devices but also active-matrix organic light-emitting diodes (AMOLEDs) can be integrated with textiles and such technology is particularly promising for display applications. Kim and Song proposed an AMOLED on textile device in 2016 [88]. Both organic thin film transistors (OTFTs) and OLEDs were fabricated on PET fabric previously planarized by using polyurethane (PU) and photo-acrylic (PA) films. All OTFT and OLED layers were evaporated and patterned by a lift-off process, except for the OTFTs' active material (specifically, TIPS-pentacene), which was inkjet-printed. As result, the OTFT mobility was 0.34 cm^2/V and the OLED luminance was 64,459 cd/m^2 at 12 V. Kim and Song recently reported another AMOLED device with improved performance [89]. A high mobility of 0.98 cm^2/V was achieved by using carbon nanotube/Au electrodes and the photo-acryl as a dielectric. Moreover, a protective layer composed of PVA and PA was used to improve the stability of devices. Indeed, after 20 days in air, the luminance of the encapsulated device decreased to 64%, while that of the bare device decreased to 54%.

In conclusion, OLEDs are well-performing light-emitting devices for textile integration. Nonetheless, some practical aspects need to be fully addressed for commercial application. Indeed, the performance and manufacturing costs of the encapsulation barrier layer represent, at the moment, the main challenge.

Table 4. Structure and performance of OLEDs on textiles.

Year	Deposition Techniques	Device Structure	Performances	Ref.
2013	Thermal evaporation	Ag/WO_3/NPB/Alq_3/Liq/Al/Ag/NPB	7000 cd/m^2 and 8 cd/A at 6 V	[77]
2014	Thermal evaporation	Al/Liq/Alq_3/NPB/WO_3/Ag	2000 cd/m^{2}at 7.5 V 3 cd/A at 6.5 V	[78]
2015	Spin coating + thermal evaporation	PEDOT:PSS/Super Yellow(PDY-132)/LiF/Al	5000 cd/m^2 at 6 V. 9.72 cd/A at 5.5 V7.17 lm/W at 4 V.	[80]
2015	thermal evaporation	Ag/HAT-CN/NPB/TAPC/CBP: Ir(ppy)$_3$/TPBi/LiF/Al	64,459 cd/m^2 at 12 V	[88]
2016	Spin coating + thermal evaporation	ITO/PEDOT:PSS/emission polymer(SPW-111, PDY-132, and SPR-001)/LiF/Al	2781cd/m^2 at 13 V 0.29 cd/A at 13 V for the white OLED; 2430cd/m^2 at 10 V 0.10 cd/A at 10 V for the red OLED; 6305 cd/m^{2}at 11 V 0.38 cd/A at 11 V for the yellow OLED	[84]
2016	Thermal evaporation	Al/Liq/Alq_3/NPB/WO_3/Ag	1500 cd/m^2 and 5 cd/A at 8.5 V	[79]
2017	Thermal evaporation	Ag/MoO_3/NPB/mCP: Ir(ppy)3(6% wt)/TPBi/Ca/Ag	45,545 cd/m^2 at 10.5 V 37.7 cd/A at 7.5 V	[85]
2017	Thermal evaporation	Al/Liq/TPBi/CBP: Ir(ppy)$_3$(8% wt)/NPB/MoO_3/Ag	93,000 cd/m^{2}at 14 V 49 cd/A at 12 V	[81]
2018	Spin coating + thermal evaporation	Ag/PEDOT:PSS/Super Yellow/Ca/Ag	n.a.	[90]
2019	Thermal evaporation	Al/Liq/$Bebq_2$:Ir(piq)$_3$/NPB/MoO_3/Ag	1660 cd/m^2 and 19 cd/Aat 8.71 mA/cm^2	[87]
2019	Thermal evaporation	Ag/HAT-CN/NPB/TAPC/CBP: Ir(ppy)$_3$/TPBi/LiF/Al/Ag	23,673 cd/m^{2}at 7 V (bare OLED)16,636 cd/m^2 at 7 V (encapsulated OLED)	[89]
2020	Thermal evaporation	Ag/MoO_3/NPB/CBP: Ir(bt)$_2$(acac)/TPBi/Ca/Ag	15,000 cd/m^{2}at 8 V 78 cd/A at 6 V.	[86]
2020	Thermal evaporation	Au/MoO_3/NPB/mCP: Ir(ppy)$_3$(6% wt)/TPBi/Ca/Ag	17,900 cd/m^2 at 10 V 12.4 cd/A at 4 V (anode side); 15,300 cd/m^2 at 10 12.8 cd/A at 4V (cathode side); total current efficiency of the transparent OLED 25.2 cd/A at 4 V	[83]
2020	Thermal evaporation + spin coating	NPB/Ag/MoO_3/NPB/Alq_3/Liq/Al ZnS/Ag/ZnO/PEI/ Super Yellow(PDY-132)/MoO_3/Ag/NPB	About 7000 cd/m^2 at 5.5 V (OLED); 10,000 cd/m^2 at 5.5 V (PLED)	[82]

2.4. LEC Devices

Light electrochemical cells (LECs) are an evolution of OLED devices; in the mid-1990s *Pei* and co-workers discovered such new technologies by modifying the emissive layer of a polymeric OLED by adding mobile ions and a solid electrolyte and they explained the charge injection from the electrodes as a redox electrochemical process [91]. In addition to conjugate polymers, this kind of phenomenon was observed also for ionic transition metal complexes, so we can distinguish between pLECs (polymeric LEC) and iTMC-LECs (ionic transition metal complexes LECs) [92]. When an external bias is applied, the electrical field induces a redistribution of the ions; in particular anions and cations migrate towards the corresponding electrodes with the formation of an electric double layer (EDL) at the interface between the electrode and the active layer. In particular, when the applied voltage is low, the carrier injection from the electrodes is limited, the ions drift is predominant and induces the formation of the EDL, while the central part of the active layer remains field-free; in these conditions, the device shows low luminance values. When the bias increases, more electrical charges are injected into the organic layer, inducing redox processes. The new charge species are stabilized by the ions located at the organic layer/electrode interface with the formation of doped zones; the result is analogous to a pin OLED structure (Figure 1d) [92]. The doping can improve the charge injection from the electrodes because it induces a band bending of the HOMO (Highest Occupied Molecular Orbital) and LUMO (Lowest Unoccupied Molecular Orbital) levels towards the Fermi level. Considering the improved charge injection, every kind of electrode can be used in LEC structure without concerns related to the work function level. The architecture of a LEC device is very simple and includes an active layer (composed of electroluminescent material, ions, and a solid electrolyte) sandwiched between two electrodes, of which at least one is transparent (Figure 1d). The active layer can be easily deposited by solution processes and higher thicknesses can be used with an improvement in the mechanical stability and less concerns related to short circuits; high stable and non-reactive electrodes can be used without degradation problems. On the other hand, if the in-situ doping process is not properly controlled, the doped regions may continue to grow towards each other until they meet at the center of the device. This effect causes a quenching process and a drastic reduction of luminance. The migration of the doped zones is strongly related to the electronic and ionic mobility of the active layer, the applied bias, and the thickness of the active layer. All of these factors should be evaluated to optimize device performances. The ionic mobility also influences the switch on voltage and time. The latter can assume values between few milliseconds to several hours and for such reason, LEC technology is not suitable for display applications.

Some attempts to integrate this technology with fabrics have been carried out in recent years, although the results are still limited and not comparable with those achieved by the other technologies already discussed. In 2012, a single electroluminescent fiber based on an iTMC was developed by using a co-electrospinning technique as reported in Figure 6a [93]. The device architecture is also illustrated in Figure 6a and consists of a Galinstan liquid metal core, an iTMC based electroluminescent layer, and an ITO coating. The fiber-based light-emitting device showed a turn-on voltage of 4.2 V, uniform light emission, and has great potential for textile integration thanks to its flexibility, conformability, and lightweight. A different approach for the fabrication of LEC devices on single fiber was developed by the group of H. Peng [94,95]. Starting from a metal wire-based cathode, ZnO and a polymeric electroluminescent layer were deposited by dip coating and finally, a carbon nanotube sheet wrapped around the coated wire was used as a transparent anode [94]. The CNT sheet was obtained by a dry-drawing process starting from a CNT array synthesized by chemical vapor deposition; it shows high transmittance and electrical conductivity around 102–103 S/cm which remain almost unvaried after 1000 bending cycles. The device reached a luminance value of about 600 cd/m^2 and an efficiency of 0.83 cd/A. The fiber-based LEC device was also fabricated by using the CNT sheet both as cathode and anode layer as you can see in Figure 6b; in this case, smaller luminance and efficiency values (505 cd/m^2 and 0.51 cd/A) were reported compared to the previous device with the CNT sheet only as a cathode [95]. A LEC device was also developed as a planar device on textile by using a spray-coating deposition [96]. The fabric was composed of conductive and polymeric fibers embedded in a polyurethane matrix and the thickness of the PU matrix was designed to have conductive bumps on the surfaces as electrical contact of the light-emitting device. On top of the textile, PEDOT:PSS, active layer, and aluminum were deposited to fabricate the device that showed high luminance (>4000 cd/m^2) and good efficiency (3.4 cd/A). Spray coating was also employed to obtain all solution-based LECs (bottom electrode—active layer—top electrode) on standard polyester cotton textile previously planarized by using a UV curable PU resin [97]. The obtained performances are poor, but this result is really interesting for the development of all solution-based devices toward a low-cost fabrication process.

As previously mentioned, the possibility to use high stable electrodes without charge injection problems is a great advantage for this kind of light-emitting architecture, compared to OLED devices. In particular, silver electrodes can be easily obtained by solution processes by using silver nanoparticles or nanowires with the possibility to fabricate all solution-based low-cost light-emitting devices on textile and other kinds of innovative substrates [98–100]. On the other hand, further efforts are needed to optimize the electro-optical characteristics of LEC devices.

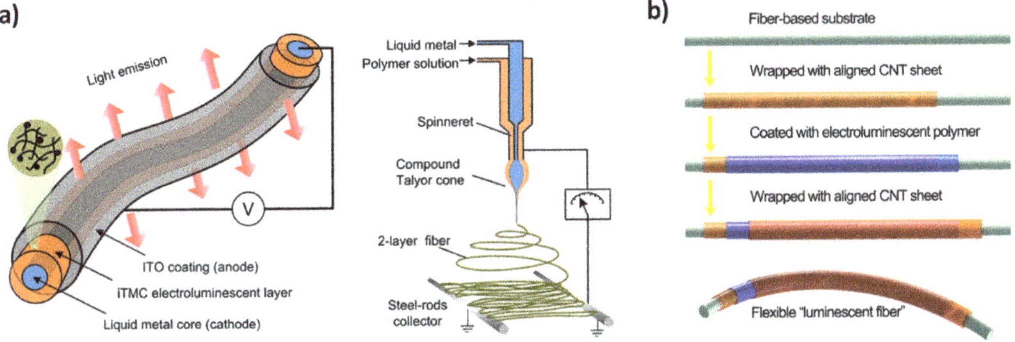

Figure 6. (**a**) Architecture of an electroluminescent fiber based on an ionic transition-metal complex (iTMCs) and scheme of the co-electrospinning setup. (**b**) Scheme of the fabrication process of a fiber-based-LEC. (**a**) Reproduced with permission [93]. Copyright 2012, American Chemical Society. (**b**) Reproduced with permission [95]. Copyright 2015, Royal Society of Chemistry.

2.5. POFs

Since the 1960s, optical fibers have been known for light transmission and communications [101]. Optical fibers are very interesting for wearable applications thanks to their similarity with traditional textile fibers, and for such reason they can be processed for fabric weaving as standard textile yarns [102]. In particular, polymeric optical fibers (POFs) are very promising for textile integration thanks to their flexibility, light weight and low cost, although they do not reach the performance levels of traditional glass fibers in terms of propagation losses and data transmission capacity [102]. A typical optical fiber is composed of a central region, the core, and by an external region, the cladding; the refractive index of the core is higher than that of the cladding. The POF is fabricated by two principal techniques: drawing form preforms and extrusion [103], and typically employed polymers for their fabrication are PMMA, polystyrene, and polycarbonates for the core, while the cladding is usually composed of fluorinated polymers [104]. A POF is a cylindrical dielectric waveguide usually employed to transport light between the two ends of the fiber by exploiting the total internal reflection phenomenon (distal-end-emitting optical fibers). Side emission can occur in specific conditions and such technology can have great opportunities for large-area lighting. Two main approaches are usually employed to obtain side emission: microperforations of the fiber or macrobending, as shown in Figure 1e [102]. Indeed, the microperforation and the damaging of the cladding induce light scattering and leakage through the cracks. Microperforation can be obtained by using different methods: notching [105,106], abrasion [105,106], laser micromachining [107,108], surface chemical etching [106,109], etc. In 2003, Harlin et al. obtained light-emitting textiles with a conventional weaving machine by using polycarbonate fibers, and different mechanical methods were tested to induce microperforations [105]. Konkar et al. compared a mechanical and a chemical treatment to obtain side-emitting POFs, and it was demonstrated that the chemical etching ensures better control [109]. In general, mechanical methods are considered more aggressive since they can cause damage to the fibers with a weakening of the core structures. To improve the robustness after mechanical treatments, polymeric overcoats can be deposited [106]. Anyway, mechanical treatments cannot be precisely controlled and they are not reproducible, as a result, the light distribution is not uniform and different from sample to sample. The laser treatment is more precise than mechanical and chemical treatments and it can also be employed to obtain specific patterns, as reported by Shen et al. [107]. Side emission from optical fibers can also be obtained via macrobending in a way to have a propagation angle higher than the critical angle and light emission out of the fiber. Different from the microperforation approach, no modification is necessary to the structure of the optical fibers, and it is sufficient to design specific embroidery or weaving architectures during the textile manufacturing to obtain side emission. In particular, the absence of post-treatment of the fiber can be considered an important aspect for the reduction of the cost. On the other hand, the intensity of the light emission in the macro-bending approach is very sensitive to the bent radius and it is necessary to ensure that a constant bending radius is maintained in the whole textile also when it is worn. For this purpose, Wang et al. reported a theoretical equation to quantify the correlation between the light emission efficiency and the bending radius of POFs [110]. Although it is essential to control the bending radius of the fiber to have a good lateral emission, it is important to avoid abrupt folding or bending of the fiber that can cause breakage and affect the light emission uniformity. To avoid the risk of breakage, weaving is usually preferred compared to knitting. Therefore, with the appropriate precautions, side-emitting POFs can be processed to form any type of fabric in combination with traditional yarns, and the result is a textile with good tactility and comfort. To obtain light-emitting textiles from lateral-emitting POFs, they are connected to external light sources together with a power source and a motherboard. LEDs are usually employed as lightweight light sources that can be connected at each end of the fiber to obtain an intense and uniform light emission. Indeed, if the LED is connected only at one end of the fiber, the light emission decreases progressively along the fiber. The motherboard can also be designed to incorporate sensors and programs to obtain different illumination and interactive effects. All these external elements are small and lightweight, and they do not affect the portability of the garments. Differently from light-emitting textiles obtained using LEDs, the light emission from POFs is not point-like and so it is possible to have a large area of emission. On the other hand, the emission area is not uniform as it is for planar light-emitting devices fabricated on textiles. Anyway, side-emitting POFs represent a robust, stable, and low-cost technology for light-emitting textiles that is already on the market.

3. Health and Environmental Concerns

The integration of electroluminescent devices with textiles raises some issues related to health and environmental concerns that should be considered for correct and safe use.

The ACEL devices are among the safest electroluminescent devices in terms of their chemical composition. Biocompatible polymers such as polyvinylpyrrolidone (PVP) and PDMS are largely employed as dielectrics in ACEL devices. As for phosphors, they are composed of a host material and a dopant. Typically, hosts are oxides, nitrides, sulfides, or selenides of zinc, cadmium, manganese, aluminum etc. Although some of these compounds can be toxic, copper-doped ZnS is the most used one, and is currently used in glow-in-the-dark toys.

Regarding LEDs, they can contain arsenic, gallium, indium, antimony etc. which can cause human and ecological toxicity effects [111]. Researchers from the University of California tested the toxicity of different kinds of LEDs by using a standardized leachability test and they observed that all the above-mentioned elements are below the threshold limit, according to California State regulations, while copper, lead, and nickel are above legal limits [112]. Such metals are usually employed in ancillary technologies such as wires, solders, glues etc., and they represent the main contribution to the total hazard potential. Moreover, such ancillary technologies are employed in all the device architectures mentioned in this review and this aspect should be considered in terms of potential toxicity.

As for OLED devices, a recent study compared the environmental impact of OLED technology with specific concern given to displays [113]. Of course, such a kind of application is quite different form the e-textile one, but it provides some indications. In particular, it was reported that silver, chromium, beryllium, and copper exceed legal limits in a leachbility test. Unfortunately, most of organic materials used in OLED and other organic-based light-emitting technologies (e.g., LECs) have not yet been tested and further studies are necessary to assess their environmental and health impacts.

No reports are available about the effect of such devices to human health. Anyway, all of them need an appropriate packaging and encapsulation to protect them from environmental agents and mechanical stress so there is no direct contact between the device and the wearer. Any effects related to the diffusion of potentially dangerous materials toward the skin could be then eliminated and/or minimized.

In light of the considerations reported, all the electroluminescent devices must be disposed in an appropriate way at the end of their life as they are potentially harmful to the environment.

4. Applications

The integration of light-emitting devices with fabrics is interesting for numerous applications in different areas. Light-emitting clothes can be used for the fashion industry and artistic exhibition but also for visual communications such as promotional events and visibility improvement in low-light conditions. Light emission gives a new look to our clothes and at the same time can be exploited to communicate with viewers. Both academia and industry are working in this application field. Shenzhen Fashion Luminous Technology Co. Ltd. is a Chinese company that was established in 2013 and it is specialized in the manufacturing of light-emitting optical-fiber-based fabrics and clothing. DreamLux is an Italian company that developed a light-emitting textile (LumIGram®) based on side-emitting optical fibers and it provides ready-for-use "Luminous Panels" composed of a piece of fabric (optical and traditional fibers), LEDs, and a power supply. Another example of light-emitting clothes was reported by Philips in 2006; Lumalive is a technology based on flexible LED arrays located beneath the outer fabric. This lighting system does not compromise the softness of the cloths and it can be easily removed when you want to wash the garments. The American startup Lumenus developed smart clothes and other accessories for runners and cyclists based on LEDs. A similar product was created by researchers at the Taiwan Textile Research Institute in collaboration with PEGA D&E and it consists of a cycling jacket integrated with LED yarns (LaightFairy). CuteCircuit is another company that has created many illuminating costumes based on LEDs. Recently, Tan explored the challenges related to the designing of POF-based textiles for the creation of dynamic fashions [114]. In the same field, researchers of electronic engineering and fashion design departments at Kookmin University published a study where they reported some technical and design guidelines for LED integration into textiles and they also developed some prototypes [115]. Hardy et al. demonstrated the possibility of integrating, via embroidering technique, a non-stretchable LED yarn with stretch fabric for a carnival costume (Figure 7d) [116]. A digital watch display [117] and a matrix display [118] on fabric were also fabricated by exploiting the ACEL architecture and screen-printing deposition technique.

Sensing is another possible application area for light-emitting textiles as both input and output elements. Textile is the closest layer to our body, and for such reason it is the ideal substrate for sensor integration for physiological monitoring; at the same time, clothes are ubiquitous in everyday life and they can be used as supports for portable environmental sensors useful to preserve our health and our quality of life. Cherenack et al. reported in 2010 on a combination of temperature and humidity sensors with LEDs as visual indicators on textile [51]. A prototype of smart clothing designed for construction worker safety was demonstrated in 2015. It consists of a temperature sensor with a light emission (LED) and an acoustic alerting system when abnormal temperatures are recorded [119]. In 2017, Liu et al. reported a textile pressure sensor for monitoring physiological signals and human motion with a real-time visual response provided by an LED [2]. LEDs have also been integrated with photodiodes for arterial oxygen saturation (Figure 7b) [120] and heart rate [121] monitoring (Figure 7a). POFs can also be used as sensors by exploiting two different approaches: wavelength modulation or intensity modulation. Bragg-grating-based fibers are usually employed for wavelength modulation. When an incident broad band light enters the gratings, a specific wavelength is reflected that is equal to $2n\Delta$, where n is the refractive index and Δ is the grating pitch [102]. As a consequence of external signals such as temperature and strains, both the n and Δ change and so the reflected wavelength and its variation can be exploited in sensing applications [122–124]. Regarding the intensity modulation, in this case the light intensity can be lost by defects and geometrical perturbation of the fiber axis and it can be related to external signals that can be monitored [124]. POF-based sensors have been largely employed for monitoring of breathing [125–128], plantar pressure and temperature monitoring [129], and monitoring of oxygen saturation of hemoglobin [130]; more details can be found in different reviews [102,104,131].

Finally, light-emitting textiles can be used for light therapy. Indeed, light has been largely applied in clinical practice for disease treatments, pain relief, tissue repair, and so on [132,133]. Textiles can improve user acceptance thanks to their conformability and the possibility to fabricate portable light-emitting fabrics. Furthermore, since the textile is very close to the skin, this reduces the amount of light that is lost and makes the treatment more effective. In 2013, a textile based on side-emitting POFs obtained via weaving was tested during in vitro experiments that demonstrated a significant increase in collagen production in human fibroblasts irradiated by the fabric [134]. Koncar worked on the development of textile light diffusers by exploiting the macro-bending of POFs with homogenous light emission and flexibility, obtaining suitable performances for photodynamic therapy [135,136], and such technology has been recently tested both during in vitro and in vivo experiments [137]. LEDs have also been integrated with textiles for light therapy; researchers from Ghent University demonstrated an LED-based textile that delivers breathability, excellent mechanical compliance, and good irradiance without excessive thermal heating [138]. All the applications illustrated here can greatly improve the quality of our lives and so all the research results will be essential for future development and improvement.

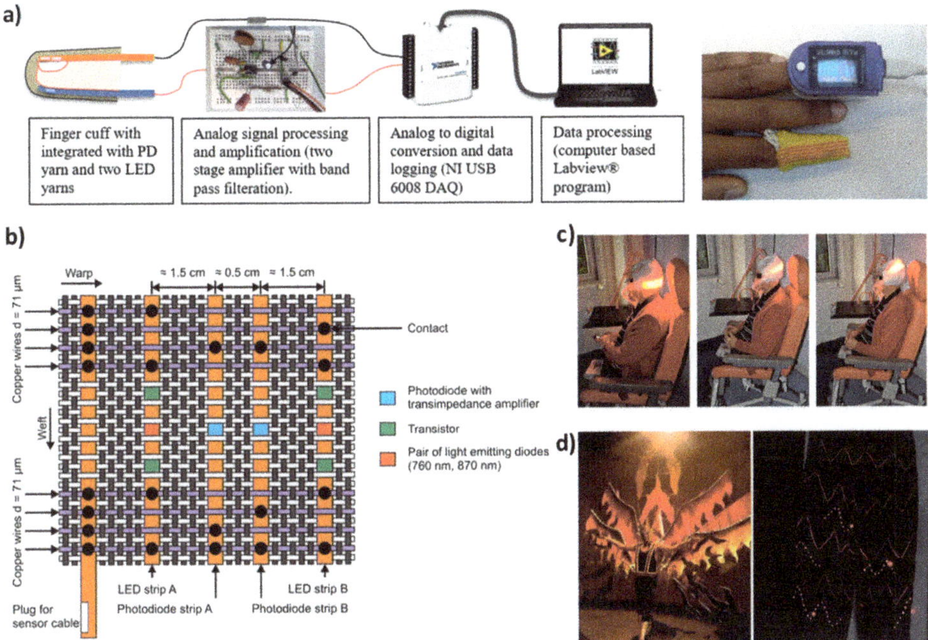

Figure 7. (**a**) Scheme of the PD/LED yarn system for heart rate detection and picture of the comparison between the e-yarn system and a commercial one. (**b**) Schematic representation of a textile-based sensor with two photodiode strips, two LED strips, and a bus bar strip, for near-infrared spectroscopy. (**c**) Picture of a photodynamic therapy session by using a POF-based light-emitting textile at 635 nm. (**d**) Carnival costume integrated with LEDs. (**a**) Reproduced under the terms of the CC BY 4.0 license [121]. (**b**) Reproduced with permission [120]. Copyright 2013, The Optical Society. (**c**) Reproduced under the terms of the CC BY 4.0 license [137]. (**d**) Reproduced under the terms of the CC BY 4.0 license [116].

5. Conclusions

Light emission is a very interesting functionality that can be integrated with textiles for applications in different fields: fashion and artistic performances, visual communications, sensing, and light therapy. Different technologies can be useful for this purpose, and they have been reported in this review. Light-emitting devices can be directly fabricated on fibers and planar textiles in the form of ACELs, OLEDs, and LECs.

Although electroluminescent fibers could in principle be employed directly by using weaving, knitting, and embroidering techniques, the fabrication on a single fiber requires the preparation of numerous devices and the design of a specific production line. Moreover, it is fundamental that the fibers are able to withstand the mechanical stress required by fabric processing. On the other hand, the manufacturing of planar electroluminescent textiles is preferable from a practical and economical point of view. Indeed, large-area planar devices can be developed directly on textiles in a single-step process, such as for devices on top of standard flexible substrates (e.g., PET). In this case, a further planarization step could be necessary to reduce surface roughness and to ensure proper operation conditions.

Of course, each device typology has positive aspects and disadvantages. ACEL devices can be easily deposited via solution-based processes by using commercially available inks. Different highly scalable deposition techniques have been developed with good results, such as screen printing, slot-die coating, dispensing printing, etc. Unfortunately, such devices show a maximum luminance of few hundreds cd/m^2 and a very high driving voltage (higher than 100 V) is usually employed; in particular, this last aspect can be considered hazardous for wearable applications.

Textile-based OLEDs show very high luminance and current efficiency and they are fabricated by using both thermal evaporation and solution-based deposition processes. However, thermal evaporation is the most used deposition technique, although it is expensive and not scalable. For such reasons, all solution-based OLED devices are a very hot topic in this field. Planarization and encapsulation represent two critical issues that must be addressed for the future development of this technology. Different kinds of planarization processes have been developed based on various polymeric materials. Encapsulation is necessary to avoid the reaction between the device and external agents (e.g., water and oxygen) that cause degradation. Typical flexible encapsulation consists of a multilayer of inorganic and organic materials deposited via atomic layer deposition (ALD) and solution processes, respectively. WVTR values close to that of glass lids (10^{-6} g/m^2/day) have been achieved on textiles with good stability results.

LECs have a device architecture similar to that of OLEDs but with two important differences: the device thickness is higher than that of OLED and so there are less concerns about short circuits; the work function of electrodes is not important for the charge injection and so high stable electrodes can be used so the entire device is also more stable. Despite these advantages over OLED devices, LECs do not show the excellent electro-optical performances reported for OLEDs, and only few examples of textile-based LECs are reported in the literature.

Inorganic LEDs are another option for the development of light-emitting textiles, but in contrast to OLEDs they are rigid and point-like devices and for such reason they are not the ideal candidates. At the same time, LEDs show very high performance and stability and this technology is commercially available. LEDs have been integrated with textiles by simply attaching them on textile-based electrical circuits or by embedding the device into the fiber, improving the mechanical robustness of the e-textile.

All the mentioned technologies consist of light-emitting devices integrated directly with textiles; however, it is possible to convey the light inside the fabric using polymeric optical fibers. Side emission can be obtained by using two different approaches, microperforation and macrobending. These fibers, coupled with LEDs, can be integrated with traditional fibers to obtain light-emitting textiles. LEDs and POFs are the most mature technologies among those mentioned in this review and they are already used in real applications. In any case, research is very active in this field and there will be important progress in the near future.

Author Contributions: Conceptualization, C.T.P., M.P. and M.C.; methodology, C.T.P., M.P. and M.C.; formal analysis, C.T.P.; investigation, C.T.P., M.C., M.P., D.T. and R.G.; data curation, C.T.P., M.C., M.P. and R.G.; writing—original draft preparation, C.T.P., M.C., M.P., R.G. and D.T.; writing—review and editing, C.T.P., M.P., R.G. and V.M.; supervision, R.G., C.T.P., V.M.; funding acquisition, V.M. and G.G. All authors have read and agreed to the published version of the manuscript.

Funding: This work was supported by: "ECO-sustainable and intelligent fibers and fabrics for TEChnic clothing (ECOTEC)", PON «R&I» 2014-2020, project N° ARS01_00951, CUP B66C18000300005; Progetto FISR—C.N.R. "Tecnopolo di nanotecnologia e fotonica per la medicina di precisione"—CUP B83B17000010001.

Institutional Review Board Statement: Not applicable.

Informed Consent Statement: Not applicable.

Data Availability Statement: Not applicable.

Conflicts of Interest: The authors declare no conflict of interest.

References

1. Stoppa, M.; Chiolerio, A. Wearable electronics and smart textiles: A critical review. *Sensors* **2014**, *14*, 11957–11992. [CrossRef]
2. Liu, M.; Pu, X.; Jiang, C.; Liu, T.; Huang, X.; Chen, L.; Du, C.; Sun, J.; Hu, W.; Wang, Z.L. Large-Area All-Textile Pressure Sensors for Monitoring Human Motion and Physiological Signals. *Adv. Mater.* **2017**, *29*, 1–9. [CrossRef]
3. Atalay, A.; Sanchez, V.; Atalay, O.; Vogt, D.M.; Haufe, F.; Wood, R.J.; Walsh, C.J. Batch Fabrication of Customizable Silicone-Textile Composite Capacitive Strain Sensors for Human Motion Tracking. *Adv. Mater. Technol.* **2017**, *2*, 1–8. [CrossRef]
4. Coppedè, N.; Tarabella, G.; Villani, M.; Calestani, D.; Iannotta, S.; Zappettini, A. Human stress monitoring through an organic cotton-fiber biosensor. *J. Mater. Chem. B* **2014**, *2*, 5620–5626. [CrossRef]
5. Zhao, Y.; Zhai, Q.; Dong, D.; An, T.; Gong, S.; Shi, Q.; Cheng, W. Highly Stretchable and Strain-Insensitive Fiber-Based Wearable Electrochemical Biosensor to Monitor Glucose in the Sweat. *Anal. Chem.* **2019**, *91*, 10–6569. [CrossRef]
6. Xiong, J.; Lee, P.S. Progress on wearable triboelectric nanogenerators in shapes of fiber, yarn, and textile. *Sci. Technol. Adv. Mater.* **2019**, *20*, 837–857. [CrossRef]
7. Wu, W.; Bai, S.; Yuan, M.; Qin, Y.; Wang, Z.L.; Jing, T. Lead zirconate titanate nanowire textile nanogenerator for wearable energy-harvesting and self-powered devices. *ACS Nano* **2012**, *6*, 6231–6235. [CrossRef]
8. Pan, S.; Yang, Z.; Chen, P.; Deng, J.; Li, H.; Peng, H. Wearable Solar Cells by Stacking Textile Electrodes. *Angew. Chem. Int. Ed.* **2014**, *53*, 6110–6114. [CrossRef]
9. Lee, S.; Lee, Y.; Park, J.; Choi, D. Stitchable organic photovoltaic cells with textile electrodes. *Nano Energy* **2014**, *9*, 88–93. [CrossRef]
10. Li, R.; Xiang, X.; Tong, X.; Zou, J.; Li, Q. Wearable Double-Twisted Fibrous Perovskite Solar Cell. *Adv. Mater.* **2015**, *27*, 3831–3835. [CrossRef]
11. Gaikwad, A.M.; Zamarayeva, A.M.; Rousseau, J.; Chu, H.; Derin, I.; Steingart, D.A. Highly Stretchable Alkaline Batteries Based on an Embedded Conductive Fabric. *Adv. Mater.* **2012**, *24*, 5071–5076. [CrossRef]
12. Pu, X.; Liu, M.; Li, L.; Han, S.; Li, X.; Jiang, C.; Du, C.; Luo, J.; Hu, W.; Wang, Z.L. Wearable Textile-Based In-Plane Microsupercapacitors. *Adv. Energy Mater.* **2016**, *6*, 1601254. [CrossRef]
13. Yang, Y.; Huang, Q.; Niu, L.; Wang, D.; Yan, C.; She, Y.; Zheng, Z. Waterproof, Ultrahigh Areal-Capacitance, Wearable Supercapacitor Fabrics. *Adv. Mater.* **2017**, *29*, 1606679. [CrossRef]
14. Lee, S.-S.; Choi, K.-H.; Kim, S.-H.; Lee, S.-Y. Wearable Supercapacitors Printed on Garments. *Adv. Funct. Mater.* **2018**, *28*, 1705571. [CrossRef]
15. Hu, L.; Pasta, M.; La Mantia, F.; Cui, L.; Jeong, S.; Deshazer, H.D.; Choi, J.W.; Han, S.M.; Cui, Y. Stretchable, porous, and conductive energy textiles. *Nano Lett.* **2010**, *10*, 708–714. [CrossRef] [PubMed]
16. Destriau, G. Recherches sur les scintillations des sulfures de zinc aux rayons α. *J. Chim. Phys.* **1936**, *33*, 587–625. [CrossRef]
17. Verboven, I.; Deferme, W. Direct printing of light-emitting devices on textile substrates. *Narrow Smart Text.* **2017**, 259–277. [CrossRef]
18. Wang, L.; Xiao, L.; Gu, H.; Sun, H. Advances in Alternating Current Electroluminescent Devices. *Adv. Opt. Mater.* **2019**, *7*, 1801154. [CrossRef]
19. Yang, Z.; Wang, W.; Pan, J.; Ye, C. Alternating Current Electroluminescent Devices with Inorganic Phosphors for Deformable Displays. *Cell Rep. Phys. Sci.* **2020**, *1*, 100213. [CrossRef]
20. Jayathilaka, W.A.D.M.; Chinnappan, A.; Tey, J.N.; Wei, J.; Ramakrishna, S. Alternative current electroluminescence and flexible light emitting devices. *J. Mater. Chem. C* **2019**, *7*, 5553–5572. [CrossRef]
21. Dias, T.; Monaragala, R. Development and analysis of novel electroluminescent yarns and fabrics for localized automotive interior illumination. *Text. Res. J.* **2012**, *82*, 1164–1176. [CrossRef]
22. Bellingham, A.A.; Fontecchio, A.A. Direct Integration of Individually Controlled Emissive Pixels into Knit Fabric for Fabric-Based Dynamic Display. *IEEE Photonics J.* **2017**, *9*, 1–10. [CrossRef]
23. Bellingham, A.; Bromhead, N.; Fontecchio, A. Rapid prototyping of slot die devices for roll to roll production of EL fibers. *Materials* **2017**, *10*, 594. [CrossRef] [PubMed]
24. Martin, A.; Fontecchio, A. Effect of fabric integration on the physical and optical performance of electroluminescent fibers for lighted textile applications. *Fibers* **2018**, *6*, 50. [CrossRef]
25. Liang, G.; Yi, M.; Hu, H.; Ding, K.; Wang, L.; Zeng, H.; Tang, J.; Liao, L.; Nan, C.; He, Y.; et al. Coaxial-Structured Weavable and Wearable Electroluminescent Fibers. *Adv. Electron. Mater.* **2017**, *3*, 1–10. [CrossRef]
26. Sun, T.; Xiu, F.; Zhou, Z.; Ban, C.; Ye, T.; Ding, Y.; Liu, J.; Huang, W. Transient fiber-shaped flexible electronics comprising dissolvable polymer composites toward multicolor lighting. *J. Mater. Chem. C* **2019**, *7*, 1472–1476. [CrossRef]
27. Hu, D.; Xu, X.; Miao, J.; Gidron, O.; Meng, H. A stretchable alternating current electroluminescent fiber. *Materials* **2018**, *11*, 184. [CrossRef]
28. Zhang, Z.; Shi, X.; Lou, H.; Cheng, X.; Xu, Y.; Zhang, J.; Li, Y.; Wang, L.; Peng, H. A one-dimensional soft and color-programmable light-emitting device. *J. Mater. Chem. C* **2018**, *6*, 1328–1333. [CrossRef]
29. Zhang, Z.; Cui, L.; Shi, X.; Tian, X.; Wang, D.; Gu, C.; Chen, E.; Cheng, X.; Xu, Y.; Hu, Y.; et al. Textile Display for Electronic and Brain-Interfaced Communications. *Adv. Mater.* **2018**, *30*, 1800323. [CrossRef]
30. Torres Alonso, E.; Rodrigues, D.P.; Khetani, M.; Shin, D.W.; De Sanctis, A.; Joulie, H.; de Schrijver, I.; Baldycheva, A.; Alves, H.; Neves, A.I.S.; et al. Graphene electronic fibres with touch-sensing and light-emitting functionalities for smart textiles. *NPJ Flex. Electron.* **2018**, *2*, 25. [CrossRef]

31. Song, S.; Song, B.; Cho, C.H.; Lim, S.K.; Jeong, S.M. Textile-fiber-embedded multiluminescent devices: A new approach to soft display systems. *Mater. Today* **2020**, *32*, 46–58. [CrossRef]
32. Wu, Y.; Mechael, S.S.; Lerma, C.; Carmichael, R.S.; Carmichael, T.B. Stretchable Ultrasheer Fabrics as Semitransparent Electrodes for Wearable Light-Emitting e-Textiles with Changeable Display Patterns. *Matter* **2020**, *2*, 882–895. [CrossRef]
33. De Vos, M.; Torah, R.; Tudor, J. Dispenser printed electroluminescent lamps on textiles for smart fabric applications. *Smart Mater. Struct.* **2016**, *25*, 045016. [CrossRef]
34. Hu, B.; Li, D.; Ala, O.; Manandhar, P.; Fan, Q.; Kasilingam, D.; Calvert, P.D. Textile-based flexible electroluminescent devices. *Adv. Funct. Mater.* **2011**, *21*, 305–311. [CrossRef]
35. Hu, B.; Li, D.; Manandharm, P.; Fan, Q.; Kasilingam, D.; Calvert, P. CNT/conducting polymer composite conductors impart high flexibility to textile electroluminescent devices. *J. Mater. Chem.* **2012**, *22*, 1598–1605. [CrossRef]
36. Sloma, M.; Janczak, D.; Wroblewski, G.; Mlozniak, A.; Jakubowska, M. Electroluminescent structures printed on paper and textile elastic substrates. *Circuit World* **2014**, *40*, 13–16. [CrossRef]
37. Janczak, D.; Zych, M.; Raczyński, T.; Dybowska-Sarapuk, Ł.; Pepłowski, A.; Krzemiński, J.; Sosna-Głębska, A.; Znajdek, K.; Sibiński, M.; Jakubowska, M. Stretchable and washable electroluminescent display screen-printed on textile. *Nanomaterials* **2019**, *9*, 1276. [CrossRef]
38. Zhang, Z.; Shi, X.; Lou, H.; Xu, Y.; Zhang, J.; Li, Y.; Cheng, X.; Peng, H. A stretchable and sensitive light-emitting fabric. *J. Mater. Chem. C* **2017**, *5*, 4139–4144. [CrossRef]
39. Wu, Y.; Mechael, S.S.; Chen, Y.; Carmichael, T.B. Solution Deposition of Conformal Gold Coatings on Knitted Fabric for E-Textiles and Electroluminescent Clothing. *Adv. Mater. Technol.* **2018**, *3*, 1–7. [CrossRef]
40. Zhou, Y.; Zhao, C.; Wang, J.; Li, Y.; Li, C.; Zhu, H.; Feng, S.; Cao, S.; Kong, D. Stretchable High-Permittivity Nanocomposites for Epidermal Alternating-Current Electroluminescent Displays. *ACS Mater. Lett.* **2019**, *1*, 511–518. [CrossRef]
41. Holonyak, N.; Bevacqua, S.F. Coherent (visible) light emission from Ga(As1-xPx) junctions. *Appl. Phys. Lett.* **1962**, *1*, 82–83. [CrossRef]
42. Guarnieri, M. An Historical Survey on Light Technologies. *IEEE Access* **2018**, *6*, 25881–25897. [CrossRef]
43. Yeh, N.G.; Wu, C.H.; Cheng, T.C. Light-emitting diodes-Their potential in biomedical applications. *Renew. Sustain. Energy Rev.* **2010**, *14*, 2161–2166. [CrossRef]
44. Buechley, L. A construction kit for electronic textiles. In Proceedings of the 10th IEEE International Symposium on Wearable Computers, Montreux, Switzerland, 11–14 October 2006; pp. 83–90. [CrossRef]
45. Linz, T.; Vieroth, R.; Dils, C.; Koch, M.; Braun, T.; Becker, K.F.; Kallmayer, C.; Hong, S.M. Embroidered interconnections and encapsulation for electronics in textiles for wearable electronics applications. *Adv. Sci. Technol.* **2008**, *60*, 85–94. [CrossRef]
46. Locher, I.; Tröster, G. Fundamental building blocks for circuits on textiles. *IEEE Trans. Adv. Packag.* **2007**, *30*, 541–550. [CrossRef]
47. Von Krshiwoblozki, M.; Linz, T.; Neudeck, A.; Kallmayer, C. Electronics in Textiles—Adhesive Bonding Technology for Reliably Embedding Electronic Modules into Textile Circuits. *Adv. Sci. Technol.* **2012**, *85*, 1–10. [CrossRef]
48. Paradiso, R.; Loriga, G.; Taccini, N. A wearable health care system based on knitted integrated sensors. *IEEE Trans. Inf. Technol. Biomed.* **2005**, *9*, 337–344. [CrossRef] [PubMed]
49. Buechley, L.; Eisenberg, M. Fabric PCBs, electronic sequins, and socket buttons: Techniques for e-textile craft. *Pers. Ubiquitous Comput.* **2009**, *13*, 133–150. [CrossRef]
50. Kim, Y.; Kim, H.; Yoo, H.J. Electrical characterization of screen-printed circuits on the fabric. *IEEE Trans. Adv. Packag.* **2010**, *33*, 196–205. [CrossRef]
51. Cherenack, K.; Zysset, C.; Kinkeldei, T.; Münzenrieder, N.; Tröster, G. Woven electronic fibers with sensing and display functions for smart textiles. *Adv. Mater.* **2010**, *22*, 5178–5182. [CrossRef]
52. Komolafe, A.; Torah, R.; Wei, Y.; Nunes-Matos, H.; Li, M.; Hardy, D.; Dias, T.; Tudor, M.; Beeby, S. Integrating Flexible Filament Circuits for E-Textile Applications. *Adv. Mater. Technol.* **2019**, *4*, 1900176. [CrossRef]
53. Linz, T.; Kallmayer, C.; Aschenbrenner, R.; Reichl, H. New Interconnection Technologies for the Integration of Electronics on Textile Substrates. In Proceedings of the Ambience 05, International Scientific Conference, Intelligent Ambience and Well-Being, Tampere, Finland, 19–20 September 2005.
54. Linz, T.; von Krshiwoblozki, M.; Walter, H.; Foerster, P. Contacting electronics to fabric circuits with nonconductive adhesive bonding. *J. Text. Inst.* **2012**, *103*, 1139–1150. [CrossRef]
55. Parkova, I.; Parkovs, I.; Vilumsone, A. Light-emitting textile display with floats for electronics covering. *Int. J. Cloth. Sci. Technol.* **2015**, *27*, 34–46. [CrossRef]
56. Parkova, I.; Viļumsone, A. Functional and aesthetic design of woven electro-textile fabrics. *J. Text. Appar. Technol. Manag.* **2013**, *8*, 3.
57. Lee, H.E.; Lee, D.; Lee, T.I.; Shin, J.H.; Choi, G.M.; Kim, C.; Lee, S.H.; Lee, J.H.; Kim, Y.H.; Kang, S.M.; et al. Wireless powered wearable micro light-emitting diodes. *Nano Energy* **2019**, *55*, 454–462. [CrossRef]
58. Nashed, M.N.; Hardy, D.A.; Hughes-Riley, T.; Dias, T. A novel method for embedding semiconductor dies within textile yarn to create electronic textiles. *Fibers* **2019**, *7*, 12. [CrossRef]
59. Rein, M.; Favrod, V.D.; Hou, C.; Khudiyev, T.; Stolyarov, A.; Cox, J.; Chung, C.C.; Chhav, C.; Ellis, M.; Joannopoulos, J.; et al. Diode fibres for fabric-based optical communications. *Nature* **2018**, *560*, 214–218. [CrossRef] [PubMed]

60. Hardy, D.A.; Anastasopoulos, I.; Nashed, M.N.; Oliveira, C.; Hughes-Riley, T.; Komolafe, A.; Tudor, J.; Torah, R.; Beeby, S.; Dias, T. Automated insertion of package dies onto wire and into a textile yarn sheath. *Microsyst. Technol.* **2019**, 1–13. [CrossRef]
61. Satharasinghe, A.; Hughes-Riley, T.; Dias, T. Photodiodes embedded within electronic textiles. *Sci. Rep.* **2018**, *8*, 16205. [CrossRef] [PubMed]
62. Tang, C.W.; Vanslyke, S.A. Organic electroluminescent diodes. *Appl. Phys. Lett.* **1987**, *51*, 913–915. [CrossRef]
63. Burroughes, J.H.; Bradley, D.D.C.; Brown, A.R.; Marks, R.N.; Mackay, K.; Friend, R.H.; Burns, P.L.; Holmes, A.B. Light-emitting diodes based on conjugated polymers. *Nature* **1990**, *347*, 539–541. [CrossRef]
64. Minaev, B.; Baryshnikov, G.; Agren, H. Principles of phosphorescent organic light emitting devices. *Phys. Chem. Chem. Phys.* **2014**, *16*, 1719–1758. [CrossRef]
65. Volz, D. Review of organic light-emitting diodes with thermally activated delayed fluorescence emitters for energy-efficient sustainable light sources and displays. *J. Photonics Energy* **2016**, *6*, 020901. [CrossRef]
66. Baldo, M.A.; O'Brien, D.F.; You, Y.; Shoustikov, A.; Sibley, S.; Thompson, M.E.; Forrest, S.R. Highly efficient phosphorescent emission from organic electroluminescent devices. *Nature* **1998**, *395*, 151–154. [CrossRef]
67. Han, T.-H.; Jeong, S.-H.; Lee, Y.; Seo, H.-K.; Kwon, S.-J.; Park, M.-H.; Lee, T.-W. Flexible transparent electrodes for organic light-emitting diodes. *J. Inf. Disp.* **2015**, *16*, 71–84. [CrossRef]
68. O'Connor, B.; An, K.H.; Zhao, Y.; Pipe, K.P.; Shtein, M. Fiber shaped organic light emitting device. *Adv. Mater.* **2007**, *19*, 3897–3900. [CrossRef]
69. Kwon, S.; Kim, W.; Kim, H.; Choi, S.; Park, B. High Luminance Fiber-Based Polymer Light-Emitting Devices by a Dip-Coating Method. *Adv. Electron. Mater.* **2015**, *1*, 1–8. [CrossRef]
70. Kwon, S.; Kim, H.; Choi, S.; Jeong, E.G.; Kim, D.; Lee, S.; Lee, H.S.; Seo, Y.C.; Choi, K.C. Weavable and Highly Efficient Organic Light-Emitting Fibers for Wearable Electronics: A Scalable, Low-Temperature Process. *Nano Lett.* **2018**, *18*, 347–356. [CrossRef]
71. Ko, K.; Lee, H.; Kim, H.M.; Lee, G.J.; Shin, S.; Kumar, N.; Song, Y.M.; Kang, J. High-performance, color-tunable fiber shaped organic light-emitting diodes. *Nanoscale* **2018**, *10*, 16184–16192. [CrossRef]
72. Ko, K.; Lee, H.; Kang, J. Flexible, Wearable Organic Light-Emitting Fibers Based on PEDOT: PSS/Ag-Fiber Embedded Hybrid Electrodes for Large-Area Textile Lighting. *Adv. Mater. Technol.* **2020**, *5*, 2000168. [CrossRef]
73. Lee, H.; Chun, Y.T. Fibertronic Quantum-Dot Light-Emitting Diode for E-Textile. *ACS Appl. Nano Mater.* **2020**, *3*, 11060–11069. [CrossRef]
74. Song, Y.J.; Kim, J.W.; Cho, H.E.; Son, Y.H.; Lee, M.H.; Lee, J.; Choi, K.C.; Lee, S.M. Fibertronic Organic Light-Emitting Diodes toward Fully Addressable, Environmentally Robust, Wearable Displays. *ACS Nano* **2020**, *14*, 1133–1140. [CrossRef]
75. Hwang, Y.H.; Kwon, S.; Shin, J.B.; Kim, H.; Son, Y.H.; Lee, H.S.; Noh, B.; Nam, M.; Choi, K.C. Bright-Multicolor, Highly Efficient, and Addressable Phosphorescent Organic Light-Emitting Fibers: Toward Wearable Textile Information Displays. *Adv. Funct. Mater.* **2021**, *31*, 2009336. [CrossRef]
76. Janietz, S.; Gruber, B.; Schattauer, S.; Schulze, K. Integration of OLEDs in Textiles. *Adv. Sci. Technol.* **2012**, *80*, 14–21. [CrossRef]
77. Kim, W.; Kwon, S.; Lee, S.M.; Kim, J.Y.; Han, Y.; Kim, E.; Choi, K.C.; Park, S.; Park, B.C. Soft fabric-based flexible organic light-emitting diodes. *Org. Electron.* **2013**, *14*, 3007–3013. [CrossRef]
78. Kim, W.; Kwon, S.; Han, Y.C.; Kim, E.; Kim, H.C.; Choi, K.C.; Kang, S.H.; Park, B.-C. 28.1: OLEDs on Textile Substrates with Planarization and Encapsulation using Multilayers for Wearable Displays. In *SID Symposium Digest of Technical Papers*; SID: Campbell, CA, USA, 2014; Volume 45, pp. 364–366. [CrossRef]
79. Kim, W.; Kwon, S.; Han, Y.C.; Kim, E.; Choi, K.C.; Kang, S.H.; Park, B.C. Reliable Actual Fabric-Based Organic Light-Emitting Diodes: Toward a Wearable Display. *Adv. Electron. Mater.* **2016**, *2*, 1–7. [CrossRef]
80. Kim, H.; Kwon, S.; Choi, S.; Choi, K.C. Solution-processed bottom-emitting polymer light- emitting diodes on a textile substrate towards a wearable display. *J. Inf. Disp.* **2015**, *16*, 179–184. [CrossRef]
81. Choi, S.; Kwon, S.; Kim, H.; Kim, W.; Kwon, J.H.; Sub, M.; Lee, H.S.; Choi, K.C. Highly Flexible and Efficient Fabric- Based Organic Light-Emitting Devices for Clothing-Shaped Wearable Displays. *Sci. Rep.* **2017**, *7*, 6424. [CrossRef] [PubMed]
82. Choi, S.; Jo, W.; Jeon, Y.; Kwon, S.; Kwon, J.H.; Son, Y.H.; Kim, J.; Park, J.H.; Kim, H.; Lee, H.S.; et al. Multi-directionally wrinkle-able textile OLEDs for clothing-type displays. *NPJ Flex. Electron.* **2020**, *4*, 1–9. [CrossRef]
83. Yin, D.; Chen, Z.Y.; Jiang, N.R.; Liu, Y.F.; Bi, Y.G.; Zhang, X.L.; Han, W.; Feng, J.; Sun, H.B. Highly transparent and flexible fabric-based organic light emitting devices for unnoticeable wearable displays. *Org. Electron.* **2020**, *76*, 105494. [CrossRef]
84. Park, M.; Lee, K.S.; Shim, J.; Liu, Y.; Lee, C.; Cho, H.; Kim, M.J.; Park, S.-J.; Yun, Y.J.; Kim, H.Y.; et al. Environment friendly, transparent nano fi ber textiles consolidated with high ef fi ciency PLEDs for wearable electronics. *Org. Electron.* **2016**, *36*, 89–96. [CrossRef]
85. Liu, Y.; An, M.; Bi, Y.; Yin, D.; Feng, J.; Sun, H. Flexible Efficient Top-Emitting Organic Light-Emitting Devices on a Silk Substrate. *IEEE Photonics J.* **2017**, *9*, 1–6. [CrossRef]
86. Yin, D.; Chen, Z.; Jiang, N.; Liu, Y.; Bi, Y.; Zhang, X. Highly Flexible Fabric-Based Organic Light-Emitting Devices for Conformal Wearable Displays. *Adv. Mater. Technol.* **2020**, *5*, 1900942. [CrossRef]
87. Jeong, E.G.; Jeon, Y.; Cho, S.H.; Choi, K.C. Textile-based washable polymer solar cells for optoelectronic modules: Toward self-powered smart clothing. *Energy Environ. Sci.* **2019**, *12*, 1878–1889. [CrossRef]

88. Kim, J.S.; Song, C.K. AMOLED panel driven by OTFTs on polyethylene fabric substrate. *Org. Electron.* **2016**, *30*, 45–51. [CrossRef]
89. Kim, J.S.; Song, C.K. Textile display with AMOLED using a stacked-pixel structure on a polyethylene terephthalate fabric substrate. *Materials* **2019**, *12*, 2000. [CrossRef]
90. Verboven, I.; Stryckers, J.; Mecnika, V.; Vandevenne, G.; Jose, M.; Deferme, W. Printing smart designs of light emitting devices with maintained textile properties. *Materials* **2018**, *11*, 290. [CrossRef]
91. Pei, Q.; Yu, G.; Zhang, C.; Yang, Y.; Heeger, A.J. Polymer Light-Emitting Electrochemical Cells. *Science* **1995**, *269*, 1086–1088. [CrossRef]
92. Meier, S.B.; Tordera, D.; Pertegás, A.; Roldán-Carmona, C.; Ortí, E.; Bolink, H.J. Light-emitting electrochemical cells: Recent progress and future prospects. *Mater. Today* **2014**, *17*, 217–223. [CrossRef]
93. Yang, H.; Lightner, C.R.; Dong, L. Light-Emitting Coaxial Nanofibers. *ACS Nano* **2012**, *6*, 622–628. [CrossRef]
94. Zhang, Z.; Guo, K.; Li, Y.; Li, X.; Guan, G.; Li, H.; Luo, Y.; Zhao, F.; Zhang, Q.; Wei, B.; et al. A colour-tunable, weavable fibre-shaped polymer light-emitting electrochemical cell. *Nat. Photonics* **2015**, *9*, 233–238. [CrossRef]
95. Zhang, Z.; Zhang, Q.; Guo, K.; Li, Y.; Li, X.; Wang, L.; Luo, Y.; Li, H.; Zhang, Y.; Guan, G.; et al. Flexible electroluminescent fiber fabricated from coaxially wound carbon nanotube sheets. *J. Mater. Chem. C* **2015**, *3*, 5621–5624. [CrossRef]
96. Lanz, T.; Sandström, A.; Tang, S.; Chabrecek, P.; Sonderegger, U.; Edman, L. A light-emission textile device: Conformal spray-sintering of a woven fabric electrode. *Flex. Print. Electron.* **2016**, *1*, 025004. [CrossRef]
97. Arumugam, S.; Li, Y.; Pearce, J.; Charlton, M.D.B.; Tudor, J.; Harrowven, D.; Beeby, S. Visible and Ultraviolet Light Emitting Electrochemical Cells Realised on Woven Textiles. *Proceedings* **2021**, *68*, 9. [CrossRef]
98. Basarir, F.; Irani, F.S.; Kosemen, A.; Camic, B.T.; Oytun, F.; Tunaboylu, B.; Shin, H.J.; Nam, K.Y.; Choi, H. Recent progresses on solution-processed silver nanowire based transparent conducting electrodes for organic solar cells. *Mater. Today Chem.* **2017**, *3*, 60–72. [CrossRef]
99. Chae, J.; Kim, H.; Youn, S.-M.; Jeong, C.; Han, E.-M.; Yun, C.; Kang, M.H. Silver-nanowire-based lamination electrode for a fully vacuum-free and solution-processed organic photovoltaic cell. *Org. Electron.* **2021**, *89*, 106046. [CrossRef]
100. Fukuda, K.; Sekine, T.; Kobayashi, Y.; Takeda, Y.; Shimizu, M.; Yamashita, N.; Kumaki, D.; Itoh, M.; Nagaoka, M.; Toda, T.; et al. Organic integrated circuits using room-temperature sintered silver nanoparticles as printed electrodes. *Org. Electron.* **2012**, *13*, 3296–3301. [CrossRef]
101. Ballato, J.; Dragic, P. Glass: The Carrier of Light—A Brief History of Optical Fiber. *Int. J. Appl. Glas. Sci.* **2016**, *7*, 413–422. [CrossRef]
102. Gong, Z.; Xiang, Z.; OuYang, X.; Zhang, J.; Lau, N.; Zhou, J.; Chan, C.C. Wearable fiber optic technology based on smart textile: A review. *Materials* **2019**, *12*, 3311. [CrossRef]
103. Bhowmik, K.; Peng, G. *Handbook of Optical Fibers*; Springer: Singapore, 2019; ISBN 978-981-10-1477-2.
104. Selm, B.; Gärel, E.A.; Rothmaier, M.; Rossi, R.M.; Scherer, L.J. Polymeric optical fiber fabrics for illumination and sensorial applications in textiles. *J. Intell. Mater. Syst. Struct.* **2010**, *21*, 1061–1071. [CrossRef]
105. Harlin, A.; Mäkinen, M.; Vuorivirta, A. Development of polymeric optical fibre fabrics as illumination elements and textile displays. *Autex Res. J.* **2003**, *3*, 1–8.
106. Im, M.H.; Park, E.J.; Kim, C.H.; Lee, M.S. Modification of plastic optical fiber for side-illumination. In *Human-Computer Interaction. Interaction Platforms and Techniques*; Lecture Notes in Computer Science; Springer: Berlin/Heidelberg, Germany, 2007; Volume 4551, pp. 1123–1129. [CrossRef]
107. Shen, J.; Tao, X.; Ying, D.; Hui, C.; Wang, G. Light-emitting fabrics integrated with structured polymer optical fibers treated with an infrared CO2 laser. *Text. Res. J.* **2013**, *83*, 730–739. [CrossRef]
108. Tan, J.; Ziqian, B. Techy Fashion: Photonic fashion design process. In Proceedings of the International Fashion Conference, Ghent, Belgium, 20–21 November 2014.
109. Koncar, V. Optical Fiber Fabric Displays. *Opt. Photonics News* **2005**, *16*, 40. [CrossRef]
110. Wang, J.; Huang, B.; Yang, B. Effect of weave structure on the side-emitting properties of polymer optical fiber jacquard fabrics. *Text. Res. J.* **2013**, *83*, 1170–1180. [CrossRef]
111. Lim, S.-R.; Schoenung, J.M. Human health and ecological toxicity potentials due to heavy metal content in waste electronic devices with flat panel displays. *J. Hazard. Mater.* **2010**, *177*, 251–259. [CrossRef] [PubMed]
112. Lim, S.R.; Kang, D.; Ogunseitan, O.A.; Schoenung, J.M. Potential environmental impacts of light-emitting diodes (LEDs): Metallic resources, toxicity, and hazardous waste classification. *Environ. Sci. Technol.* **2011**, *45*, 320–327. [CrossRef]
113. Yeom, J.M.; Jung, H.J.; Choi, S.Y.; Lee, D.S.; Lim, S.R. Environmental Effects of the Technology Transition from Liquid-Crystal Display (LCD) to Organic Light-Emitting Diode (OLED) Display from an E-Waste Management Perspective. *Int. J. Environ. Res.* **2018**, *12*, 479–488. [CrossRef]
114. Tan, J. Photonic Patterns: Fashion Cutting with Illuminating Polymeric Optical Fibre (POF) Textiles. In Proceedings of the SecondInternational Conference for Creative Pattern Cutting, Huddersfield, UK, 24–25 February 2016.
115. Kim, J.E.; Kim, Y.H.; Oh, J.H.; Kim, K.D. Interactive smart fashion using user-oriented visible light communication: The case of modular strapped cuffs and zipper slider types. *Wirel. Commun. Mob. Comput.* **2017**, *2017*, 5203053. [CrossRef]
116. Hardy, D.A.; Moneta, A.; Sakalyte, V.; Connolly, L.; Shahidi, A.; Hughes-Riley, T. Engineering a costume for performance using illuminated LED-yarns. *Fibers* **2018**, *6*, 35. [CrossRef]

117. De Vos, M.; Torah, R.; Glanc-Gostkiewicz, M.; Tudor, J. A Complex Multilayer Screen-Printed Electroluminescent Watch Display on Fabric. *J. Disp. Technol.* **2016**, *12*, 1757–1763. [CrossRef]
118. Ivanov, A.; Wurzer, M. Integration of screen-printed electroluminescent matrix displays in smart textile items—Implementation and evaluation. In Proceedings of the EMPC 2017—21st European Microelectronics Packaging Conference & Exhibition, Warsaw, Poland, 10–13 September 2017; pp. 1–6. [CrossRef]
119. Edirisinghe, R.; Blismas, N. A prototype of smart clothing for construction work health and safety. In Proceedings of the CIB W099 International Health and Safety Conference: Benefitting Workers and Society through Inherently Safe Construction, Jordanstown, Northern Ireland, 10–11 September 2015; pp. 1–11.
120. Zysset, C.; Nasseri, N.; Büthe, L.; Münzenrieder, N.; Kinkeldei, T.; Petti, L.; Kleiser, S.; Salvatore, G.A.; Wolf, M.; Tröster, G. Textile integrated sensors and actuators for near-infrared spectroscopy. *Opt. Express* **2013**, *21*, 3213. [CrossRef]
121. Satharasinghe, A.; Hughes-Riley, T.; Dias, T. Photodiode and LED embedded textiles for wearable healthcare applications. In Proceedings of the 19th World Textile Conference: Textiles at the Crossroads, Ghent, Belgium, 11–15 June 2019; pp. 11–15.
122. Melle, S.M.; Liu, K.; Measures, R.M. Practical fiber-optic Bragg grating strain gauge system. *Appl. Opt.* **1993**, *32*, 3601–3609. [CrossRef] [PubMed]
123. Zhang, W.; Webb, D.J.; Peng, G.-D. Enhancing the sensitivity of poly(methyl methacrylate) based optical fiber Bragg grating temperature sensors. *Opt. Lett.* **2015**, *40*, 4046–4049. [CrossRef] [PubMed]
124. Ramakrishnan, M.; Rajan, G.; Semenova, Y.; Farrell, G. Overview of Fiber Optic Sensor Technologies for Strain/Temperature Sensing Applications in Composite Materials. *Sensors* **2016**, *16*, 99. [CrossRef] [PubMed]
125. Li, J.; Chen, J.; Xu, F. Sensitive and Wearable Optical Microfiber Sensor for Human Health Monitoring. *Adv. Mater. Technol.* **2018**, *3*, 1800296. [CrossRef]
126. Hu, H.; Sun, S.; Lv, R.; Zhao, Y. Design and experiment of an optical fiber micro bend sensor for respiration monitoring. *Sensors Actuators A Phys.* **2016**, *251*, 126–133. [CrossRef]
127. Ciocchetti, M.; Massaroni, C.; Saccomandi, P.; Caponero, M.A.; Polimadei, A.; Formica, D.; Schena, E. Smart Textile Based on Fiber Bragg Grating Sensors for Respiratory Monitoring: Design and Preliminary Trials. *Biosensors* **2015**, *5*, 602–615. [CrossRef] [PubMed]
128. Koyama, Y.; Nishiyama, M.; Watanabe, K. Smart Textile Using Hetero-Core Optical Fiber for Heartbeat and Respiration Monitoring. *IEEE Sens. J.* **2018**, *18*, 6175–6180. [CrossRef]
129. Najafi, B.; Mohseni, H.; Grewal, G.S.; Talal, T.K.; Menzies, R.A.; Armstrong, D.G. An Optical-Fiber-Based Smart Textile (Smart Socks) to Manage Biomechanical Risk Factors Associated with Diabetic Foot Amputation. *J. Diabetes Sci. Technol.* **2017**, *11*, 668–677. [CrossRef]
130. Rothmaier, M.; Selm, B.; Spichtig, S.; Haensse, D.; Wolf, M. Photonic textiles for pulse oximetry. *Opt. Express* **2008**, *16*, 12973. [CrossRef]
131. Quandt, B.M.; Scherer, L.J.; Boesel, L.F.; Wolf, M.; Bona, G.L.; Rossi, R.M. Body-Monitoring and Health Supervision by Means of Optical Fiber-Based Sensing Systems in Medical Textiles. *Adv. Healthc. Mater.* **2015**, *4*, 330–355. [CrossRef] [PubMed]
132. Yun, S.H.; Kwok, S.J.J. Light in diagnosis, therapy and surgery. *Nat. Biomed. Eng.* **2017**, *1*, 8. [CrossRef] [PubMed]
133. Ferraresi, C.; Parizotto, N.A.; Pires de Sousa, M.V.; Kaippert, B.; Huang, Y.-Y.; Koiso, T.; Bagnato, V.S.; Hamblin, M.R. Light-emitting diode therapy in exercise-trained mice increases muscle performance, cytochrome c oxidase activity, ATP and cell proliferation. *J. Biophotonics* **2015**, *8*, 740–754. [CrossRef] [PubMed]
134. Shen, J.; Chui, C.; Tao, X. Luminous fabric devices for wearable low-level light therapy. *Biomed. Opt. Express* **2013**, *4*, 2925. [CrossRef]
135. Cochrane, C.; Mordon, S.R.; Lesage, J.C.; Koncar, V. New design of textile light diffusers for photodynamic therapy. *Mater. Sci. Eng. C* **2013**, *33*, 1170–1175. [CrossRef]
136. Mordon, S.; Cochrane, C.; Tylcz, J.B.; Betrouni, N.; Mortier, L.; Koncar, V. Light emitting fabric technologies for photodynamic therapy. *Photodiagnosis Photodyn. Ther.* **2015**, *12*, 1–8. [CrossRef] [PubMed]
137. Mordon, S.; Thécua, E.; Ziane, L.; Lecomte, F.; Deleporte, P.; Baert, G.; Vignion-Dewalle, A. Light emitting fabrics for photodynamic therapy: Technology, experimental and clinical applications. *Transl. Biophotonics* **2020**, *2*, 1–10. [CrossRef]
138. Jablonski, M.; Bossuyt, F.; Vanfleteren, J.; Vervust, T.; De Smet, H. Conformable, Low Level Light Therapy platform. In *Biophotonics: Photonic Solutions for Better Health Care IV*; SPIE: Washington, DC, USA, 2014; Volume 9129, p. 912921. [CrossRef]

 micromachines

Review

Tissue Engineering Meets Nanotechnology: Molecular Mechanism Modulations in Cornea Regeneration

Olja Mijanović [1,*], Timofey Pylaev [2,3], Angelina Nikitkina [1], Margarita Artyukhova [1], Ana Branković [4], Maria Peshkova [1,5], Polina Bikmulina [1,5], Boris Turk [1,6,7], Sergey Bolevich [8], Sergei Avetisov [9,10] and Peter Timashev [1,5,11]

1 Institute for Regenerative Medicine, Sechenov University, 8-2 Trubetskaya St., 119991 Moscow, Russia; nikitkinaangelina@gmail.com (A.N.); margarita.artyukhova@gmail.com (M.A.); peshkova_m_a@staff.sechenov.ru (M.P.); bikmulina_p_yu@staff.sechenov.ru (P.B.); boris.turk@ijs.si (B.T.); timashev_p_s@staff.sechenov.ru (P.T.)
2 Saratov Medical State University N.A. V.I. Razumovsky, 112 Bolshaya Kazachya St., 410012 Saratov, Russia; pylaev_t@ibppm.ru
3 Institute of Biochemistry and Physiology of Plants and Microorganisms, Russian Academy of Sciences, 13 Prospekt Entuziastov, 410049 Saratov, Russia
4 Department of Forensic Engineering, University of Criminal Investigation and Police Studies, 196 Cara Dušana St., Belgrade 11000, Serbia; ana.brankovic@kpu.edu.rs
5 World-Class Research Center "Digital biodesign and personalized healthcare", Sechenov University, 8-2 Trubetskaya St., 119991 Moscow, Russia
6 Department of Biochemistry and Molecular and Structural Biology, Jožef Stefan Institute, 1000 Ljubljana, Slovenia
7 Faculty of Chemistry and Chemical Technology, University of Ljubljana, 1000 Ljubljana, Slovenia
8 Department of Human Pathology, Sechenov University, 8-2 Trubetskaya St., 119991 Moscow, Russia; bolevich2011@yandex.ru
9 Department of Eye Diseases, Sechenov University, 8-2 Trubetskaya St., 119991 Moscow, Russia; s.avetisov@niigb.ru
10 Research Institute of Eye Diseases, 11 Rossolimo St., 119021 Moscow, Russia
11 Chemistry Department, Lomonosov Moscow State University, Leninskiye Gory 1-3, 119991 Moscow, Russia
* Correspondence: olja.mijanovic@gmail.com

Citation: Mijanović, O.; Pylaev, T.; Nikitkina, A.; Artyukhova, M.; Branković, A.; Peshkova, M.; Bikmulina, P.; Turk, B.; Bolevich, S.; Avetisov, S.; et al. Tissue Engineering Meets Nanotechnology: Molecular Mechanism Modulations in Cornea Regeneration. *Micromachines* **2021**, *12*, 1336. https://doi.org/10.3390/mi12111336

Academic Editors: Maria Serena Chiriacò, Francesco Ferrara and Elisabetta Primiceri

Received: 23 April 2021
Accepted: 14 October 2021
Published: 30 October 2021

Abstract: Nowadays, tissue engineering is one of the most promising approaches for the regeneration of various tissues and organs, including the cornea. However, the inability of biomaterial scaffolds to successfully integrate into the environment of surrounding tissues is one of the main challenges that sufficiently limits the restoration of damaged corneal tissues. Thus, the modulation of molecular and cellular mechanisms is important and necessary for successful graft integration and long-term survival. The dynamics of molecular interactions affecting the site of injury will determine the corneal transplantation efficacy and the post-surgery clinical outcome. The interactions between biomaterial surfaces, cells and their microenvironment can regulate cell behavior and alter their physiology and signaling pathways. Nanotechnology is an advantageous tool for the current understanding, coordination, and directed regulation of molecular cell–transplant interactions on behalf of the healing of corneal wounds. Therefore, the use of various nanotechnological strategies will provide new solutions to the problem of corneal allograft rejection, by modulating and regulating host–graft interaction dynamics towards proper integration and long-term functionality of the transplant.

Keywords: cornea regeneration; tissue engineering; nanotechnology; molecular mechanisms

1. Introduction

The human cornea is a complex five-layer structure that both protects the eye and refracts light, contributing greatly to the eye's optical power. Proper light refraction is ensured through collagen fibrils' special organization in the three inner layers of the cornea [1]. For example, in Bowman's membrane collagen fibrils are randomly organized and tightly woven; at the same time, they form multiple lamellae in the stroma and hexagonal lattice in Descemet's membrane [2]. This unique microarchitecture, on the one hand, maintains corneal shape, and on the other hand ensures transparency, these two factors being essential for proper light refraction [1].

Various mechanical, chemical or thermal traumatic factors can impair cornea integrity and homogeneity [1]. The corneal epithelium, the outermost layer of the cornea, undergoes constant self-renewal due to the proliferation and migration of populations of the progenitor limbus cells located at the cornea and sclera border [3], and has good regenerative properties, allowing self-healing of superficial corneal injuries. However, deeper corneal damage can lead to severe vision impairment, requiring a corneal transplant.

Currently, the only approved treatment strategy for corneal damage is keratoplasty [2,3]. However, the lack of donor tissue, transplant rejection, and various complications following the surgery significantly reduce the effectiveness of the procedure [4,5]. In recent years, in order to solve the problem of cornea allograft deficiency, attention had been drawn to artificial cornea manufacturing using various tissue engineering approaches [6].

Despite significant advances, the engineering of corneal tissue still faces many limitations and has a number of disadvantages [7]. One of the main limitations of artificial corneal transplants is poor integration with native host tissues and high rejection rates due to various immune responses in native corneal tissues, which are especially acute in the damaged area [8]. The problems of interactions between a tissue-engineered (TE) graft and the host microenvironment will be further discussed in this review, as well as the recent studies on the multiple molecular mechanisms modulating the processes of healing, inflammation, and remodeling of the extracellular matrix (ECM) [9] and cell death [10–12].

State-of-the-art nanotechnology-based methods seem to be able to help in modulating these molecular mechanisms that are crucial for successful graft integration, thus helping to overcome some existing drawbacks in corneal tissue engineering. Current research on corneal regeneration by the means of nanotechnology and nanomedicine implies the use of nanocarriers for drug delivery, gene therapy agents [13], etc., as well as the use of nanostructured matrices to improve cell adhesion and proliferation. Many studies were devoted to the development of effective strategies for intracorneal nanomaterial (NM)-based delivery of various biomolecules, such as DNA [14], antibodies [15], peptides [16], and therapeutic agents [17]. In addition, extensive research has been conducted in the field of "smart" biocompatible nanoscaffolds, implying synthetic and semi-synthetic biomaterials [18] with desired properties and customizable structures designed for specific tasks.

Modulating molecular interactions between native cells and the transplanted biomaterial through various physicochemical [19] and nanotechnological [20] approaches will help to guide the healing processes in the damaged corneal tissues and improve clinical outcomes in patients with corneal injuries and diseases.

2. Corneal Tissue Engineering

With the background of restricted possibilities of traditional methods for cornea regeneration, transplantation, and hardly accessible donor cornea, novel methods are in high demand. The implementation of biomaterials opens a new field in approaches to cornea regeneration.

2.1. Different Approaches for Cornea Replacement

The widespread technique of organ replacement with xenogenic, decellularized, ECM-enriched matrices allows for the preservation of the native composition and anatomy of the target tissue. Such matrices usually do not contain any cells, but support the regeneration of host cells. The mechanical and optical properties of the decellularized cornea (DC) are similar to the native one [21,22]. Another example of a decellularized matrix for cornea regeneration is an amniotic membrane (AM). An AM has a three-layered structure that resembles the corneal epithelium structure and has great regenerative and anti-inflammatory properties [23]. However, both DC and AM have strong disadvantages as a cornea analog. These include the possibility of a strong immune response to remaining collagen fibrils, the possible transmission of infectious diseases, changes in stroma structure, transparency and biomechanical stability during the material preparation, and lack of complete re-cellularization after implantation [23–27].

The polar opposite approach implies the creation of a native-resembling environment de novo. Unique properties of either synthetic or natural biomaterials often provide tunability required for cornea regeneration. For instance, fibrin glue is used to restore corneal integrity after frequent intraoperative and postoperative corneal traumas and perforations [28,29]. A chemically modified UV crosslinkable material based on GelCORE gelatin has been developed which mimics the natural stiffness of the cornea and is highly adhesive, cytocompatible, and biodegradable. The hydrogel was able to seal corneal defects without the need for suturing and promoted re-epithelialization of the corneal surface [30]. Clinically available synthetic corneas are widely used to replace donor ones, including keratoprosthesis (KPro made from poly (methyl methacrylate) (PMMA) [31] and AlphaCor^{TM} poly (2-hydroxyethyl methacrylate) (PHEMA) [32]. However, natural biomaterials such as fibrin and gelatin often face rapid degradation rates, weak mechanical properties, and can act as a physical barrier for the migrating epithelial cells [33,34]. Serious side effects of artificial corneas based on synthetic matrices were reported, among them an acute foreign body response and hyperacidity of degradation products, which lead to corneal scarring [35].

To overcome existing obstacles, numerous approaches for the modification of biomaterials can be applied. ECM-containing matrices can be "strengthened" by crosslinking [36] or by combining with other materials, such as PCL nanofibers [37]. To improve biocompatibility, synthetic materials are also combined with natural biopolymers, e.g., PCL combined with collagen, gelatin, or chitosan [35,38–40]. Therefore, the combination of various biomaterials is the most promising strategy for cornea regeneration in terms of biocompatibility, mechanical properties, transparency, immune response, and cell behavior.

The tissue engineering (TE) approach satisfies these inquiries perfectly since it combines biomaterials, biochemical factors, and cells to form tissue-like structures. Corneal TE has attracted great interest recently due to avoiding many of the complications encountered in traditional donor corneal transplantation. The involvement of cells allows not only the creation of a cornea analog, but also give promises for the full regeneration and integration of the graft.

2.2. *Cornea TE Grafts*

2.2.1. Hydrogel-Based Grafts

Since collagen is the major protein in the cornea ECM, it is widely used both as a base and in addition to hydrogel grafts for cornea regeneration. Collagen vitrigel is widely used for the construction of corneal equivalents [41] and the treatment of corneal endothelial dysfunction [42]. Wang et al. cultured primary human corneal endothelial cells (HCECs), exhibited elongated morphology, and increased expression of corneal endothelial markers ZO-1 and Na^+/K^+ -ATPase on a collagen vitrigel [35]. Crosslinking collagen, e.g., with riboflavin (RF), is widely used to significantly improve its mechanical stiffness and chemical stability [19,43–45]. Fang Chen et al. [46] stitched collagen and HA together directly in a rabbit cornea wound in situ without a catalyst or light activation. The growth of corneal epithelial cells on the gel surface was maintained for 7 days, and no inflammation was found in the surrounding tissue [46].

Gelatine-based materials often have good transparency due to their high water content, which makes them a promising candidate for use in ocular tissue engineering [47]. Goodarzi et al. [48] studied the possibility of using a hydrogel based on type I collagen and crosslinked EDC/NHS gelatin, as an equivalent of the cornea, in which the MSCs of human bone marrow were encapsulated. The results show that the inclusion of COL-I increases optical properties, hydrophilicity, rigidity, and Young's modulus.

An alginate-chitosan hydrogel was created for the transplantation of LSC cells for corneal reconstruction after alkaline corneal burns. LSC cells cultured in vitro expressed stemness marker p63, but did not show expression of differentiating epithelial markers of cytokeratin 3 and 12. However, a significant improvement in epithelial repair was shown [49]. To enhance the mechanical properties of alginate, PCL matrices and PCL/chitosan electrospinning matrices were incorporated into alginate hydrogels for the treatment of corneal lesions [50].

Thermosensitive hydrogels based on chitosan were demonstrated as a promising treatment strategy for alkaline corneal burns. This approach reduced the inflammatory and apoptotic processes in the damaged corneal tissues [51]. The inclusion of stromal cell factor-1 alpha (SDF-1 alpha) in a thermosensitive chitosan-gelatin hydrogel improved the regeneration of the epithelium of the corneal damaged by alkali; LESCs expressed the characteristic marker ΔNp63. The formation of a dense epithelium occurred due to stem cell homing and the secretion of growth factors through the axis of chemokines SDF-1/CXCR4 [52].

Hyaluronic acid (HA) can be a favorable addition to hydrogel composition due to its viscosity, biocompatibility, biodegradability, non-toxicity, and significant mucoadhesive properties [53]. Moreover, HA suppresses the expression of inflammatory cytokines and increases the expression of anti-inflammatory cytokines associated with tissue repair and healing [54]. The negative charge of HA promotes adhesion on the ocular surface, contributing to a longer therapeutic effect and allowing drug molecules to permeate the cells of the mucous epithelium [53].

2.2.2. Membrane- and Film-Based Grafts

Another strategy for cornea TE is utilizing flat surfaces for reducing the graft thickness, which is crucial for the specific structure of the cornea. The fibroin membrane is able to support the formation of a multi-layered epithelium and the growth of human corneal limbal epithelial (hCLE) cells, and is currently considered a standard substrate used for corneal epithelial cell transplantation [55]. When hybrid films based on tropoelastin were constructed, the obtained membranes were optically transparent, permeable to glucose, and also supported the growth and function of epithelial and endothelial cells [56]. To increase the transparency of the corneal equivalent, hydroxypropyl methylcellulose (HPMC) was introduced into collagen to create transparent matrices with a high rate of light transmission [45]. The permeability for glucose, tryptophan, and NaCl was high in such membranes and similar to the native human cornea. This membrane supported the adhesion and proliferation of human corneal epithelial cells (HCECs). Seven months after the implantation of collagen-HPMC membranes into the cornea of rabbits, high optical transparency and growth of stromal keratocytes were maintained [45]. Wenhua Xu et al. [57] developed a membrane based on carboxymethylchitosan, hyaluronic acid, and gelatin as a carrier for primary rabbit corneal epithelial cells (CEpCs). A cell construct has been used to treat alkali-induced corneal damage in rabbits. The resulting membrane was found to be transparent, biodegradable, and suitable for CEpC attachment and proliferation. As noted above AM also showed potential as a mechanical support of transplanted limbal stem cells (LSCs) [37], limbal epithelial stem cells (LESCs) [58], corneal endothelial cells (CEnCs) [59], CSC cells [60], and CEC cells [61].

2.3. *Modulation of Cell Behavior by Cornea TE Grafts*

Newer studies have shown that neighboring mechanical surroundings (the structure, composition, and compliance of the extracellular matrix) strongly influence the behavior of LESCs. Recently, it was shown that the efficiency of cell differentiation probably depends on the biomaterials used and the composition of the cultured medium. The closer the environment resembles the human cornea, the higher the likelihood that most of the MSCs differentiate into corneal keratocytes [62]. Therefore, some authors have proposed the modulation of tissue biomechanics (e.g., substrate stiffness) as a pharmacological method for modulating the phenotype of cells, initiating novel prospects for establishing more successful cell therapies and medical devices for cornea tissue regeneration [63]. For instance, fiber orientation and composition have a significant impact on the behavior of cells in the scaffold [64]. Julia Fernández-Pérez et al. [64] obtained matrices based on decellularized corneal ECM and PCL by electrospinning to mimic the fibrous structure of the cornea. Fiber alignment and ECM incorporation influenced cell morphology and migration but did not significantly affect the phenotype. Keratocyte markers were increased in all types of scaffolds compared to TCPS.

Therefore, the TE approach provides plenty of cues vital for the establishment of the full-fledged cornea analog. Among them are high optical transparency, mechanical integrity, and proper cell behavior, leading to effective re-epithelialization. However, biocompatibility with host tissues remains crucial for preventing infection and promoting implant integration [65].

3. Molecular Pathways and Interactions between Host Tissues and a Graft

In response to biomaterial implantation, cascades of molecular pathways are triggered. They determine the success of graft integration as well as its biological activity and functionality. Degradation products released by TE scaffolds, as well as subsequent changes in the characteristics of the biomaterial surface, activate an immune response in the host tissues [66]. Biomaterials frequently induce adverse immune responses in host tissues, which result in extensive inflammatory reactions, impaired healing processes, fibrous encapsulation, and rejection of the TE construct [67]. The main strategies for healing and launching regenerative processes in corneal cells have focused on such areas as modulation of the immune response, prevention of angiogenesis, and the modulation of cell interactions [2]. In order to develop efficient strategies to overcome the undesirable biological interactions with transplanted biomaterial, a deeper understanding of the interplay between the transplant and the native host environment, as well as the damage and graft-induced alterations in molecular signaling, is required.

3.1. Processes Involved in Cornea Healing

3.1.1. ECM Reorganization and Re-Epithelization

The process of corneal wound healing is regulated by the interplay between the corneal epithelium, the Bowman layer, and the corneal stroma. An important role in this process is played by ECM and dissolvable factors produced by corneal epithelial cells and keratocytes [68]. Damage to epithelial cells may result in pathological ECM reorganization. Keratocytes around the site of injury trigger apoptosis and many of them transdifferentiate into fibroblasts. Some fibroblasts will produce α-SMA and become myofibroblasts under the influence of TGF-β and other soluble factors [69]. These non-transparent cells produce large amounts of disorganized ECM in the anterior part of the stroma, eventually hazing it and leading to a loss of corneal transparency [70]. Dysfunction in a group of signaling transduction pathways, e.g., Wnt signaling pathway (or JAK/STAT, MAPK, and PI3K/Akt signaling pathways), triggers the pathological transdifferentiation of a corneal epithelium into a skin-like epithelium [71], which results in impaired corneal regeneration. Guo et al. discovered that miR-10b (the Wnt signaling pathway) and three intersection genes (dedicator of cytokinesis 9, neuronal differentiation 1, and activated leukocyte cell adhesion molecule) may cooperate and play a key role in the process of transdifferentiation. The changes in ECM organization are perceived by transmembrane surface proteins, such as integrins, that result in the activation of various intracellular signaling cascades, mainly the focal adhesion kinase (FAK)–Src complex [72]. Activation of the FAK–Src pathway leads to re-epithelialization of the injured tissue. A sharp increase in the expression of matrix metalloproteinases (MMPs) and proteases is observed in the process of corneal wound healing. MMPs are also associated with the degradation of type I, II, and III collagen, a major ECM component. The expression of MMPs in the cornea is modulated by cytokines (such as IL-1b and IL-6) and growth factors (such as TGF-β) through tuning the expression of several transcription factors, such as AP-1 and Sp1 [73,74].

3.1.2. Soluble Factors

Growth factors (GF) play a pivotal role in corneal regeneration. Platelet-derived GF (PDGF), transforming GF beta (TGF-β), and hepatocyte GF (HGF) were shown to play a key role in modulating cell proliferation and myofibroblast differentiation [2]. Studies have indicated that HGF promotes the proliferation of CECs. Moreover, HGF treatment reversed the antiproliferative effect of IL-1β in vitro, indicating that HGF actively suppressed the inflammatory environment in the corneal epithelium. On the other hand, HGF significantly reduced the infiltration of DC45+ inflammatory cells in the cornea [2,75]. Salabarria et al. showed that local VEGFR1/R2 trap treatment prior to transplantation increases transplantation success. This treatment suppresses corneal tissue infiltration with CD11c+ dendritic cells and stimulates the local expression of pro-inflammatory and immune-regulatory cytokines [76].

3.1.3. Oxidative Stress

Endothelial cell loss after corneal transplantation may be caused by oxidative stress and endoplasmic reticulum (ER) stress [12]. The mechanism of oxidative-stress-induced apoptosis starts when inflammatory cytokines promote the production of reactive oxygen species which set off permeabilization of the mitochondrial membrane as well as the NF-κB signaling pathway [12]. NF-κB signaling pathway activation stimulates the aging of vascular endothelial cells. The ER stress mechanism is also triggered by cytokines. It causes apoptosis through the TGF-β signaling pathway [12].

3.2. Modulation of Cornea Regeneration by Biomaterials

3.2.1. Re-Epithelization

Recent studies show that mechanical properties, including rigidity, stiffness, and elasticity, affect cell behavior, as well as their ability to adhere, proliferate, and differentiate [64]. The orientation of biomaterial fibers and their composition also have a significant impact on the biocompatibility, inflammation, neovascularization, and cell behavior on the scaffold [77,78].

Rapid re-epithelialization is critical to prevent infection and promote implant/host integration. Previously described in vitro studies have shown that biomaterial mechanics and surface roughness affect the migration and maturation of epithelial cells [63]. To address the problem of re-epithelialization, Wang et al. added a thin, structurally uniform biosynthetic Bowman membrane of non-lamellar amorphous collagen I over the collagen layer of the corneal stroma to create a bilayer equivalent of the cornea. Epithelial cells formed multilayer structures on top of sBM and expressed key markers of limbal stem cells and epithelial cells p63, K3, K12, K14, and tight junction protein ZO-1 [12].

3.2.2. ECM Analogs

Corneal cells actively interact with implanted biomaterials. For instance, ECM adhesion proteins, such as fibronectin and vitronectin, adhere to the surface of the biomaterial and play an important role in modulating the inflammatory response to the biomaterial [79]. Fibronectin and vintronectin enhance cell adhesion [80,81], promote macrophage fusion, and participate in the chronic phase of a foreign body response (FBR) [82,83].

3.2.3. Mechanical Properties

The stiffness of the TE corneal constructs affects cellular spatial migration and the phenotype. This downstream signaling includes RhoA and Rho-kinase proteins that modulate the cytoskeletal structure by inducing contractility or migration through actin and myosin [84]. Some authors suggested that the modulation of tissue biomechanics may present a controlling mechanism for pharmacological control of CEC phenotypes [63].

Gouveia et al. demonstrated that soft substrates, similar to the limbus, stimulate cell proliferation and stratification without influencing cell survival. They proposed that soft substrates induce YAP inactivation and keep ΔNp63, β-catenin, and ABCG2 expression levels high. ΔNp63 inhibits YAP and Wnt/β-catenin signaling and, at the same time, activates Sox9, which enhances the expression of stem cell markers such as ABCG2 and CK15. Next, β-catenin promotes pro-proliferation factors (e.g., Ki67, cyclin D1, and Myc) and inhibits BMP4 expression. As stratification progresses, the role of soft substrates decreases and YAP activates, leading to cell differentiation [85].

3.2.4. Surface Properties and Topography

Interaction of the biomaterial surface with adsorbed proteins is crucial in the immune response to the implant [86]. Various methods of altering surface chemistry have been tested to create poorly adhesive surfaces in order to control the amount, composition, and conformational changes of bounded proteins [87]. The immune system has developed the ability to recognize hydrophobic components in biomolecules as a universal molecular pattern associated with damage, thereby triggering pattern recognition receptors and leading to biological elimination [88]. The average unfolding of a protein molecule [89] and total spreading [90] are greater on hydrophobic than on hydrophilic surfaces, where proteins retain their inherent secondary structure and show little or no adsorption on the biomaterial surface [91]. To neutralize the immunogenic effects of hydrophobic surfaces, scaffolds can be modified with hydrophilic molecules such as poly(ethylene oxide) (PEO) and PEG [79]. Additionally, the surface chemistry of a biomaterial can be changed by attaching hydrophilic functional groups such as -COOH, -OH, or -NH2, allowing the regulation of protein adsorption, complement activation, and immune cell adhesion on the surface of the material [92]. Recently, researchers succeeded in the preservation of the native 3D conformation (since unfolding or misfolding of the protein molecule itself can cause adverse reactions) instead of excluding any interaction of the graft with the surrounding tissue [93].

A surface charge is another important modulator of the host immune response. Positively charged particles promote extensive activation of the inflammatory cascades, while negatively charged surfaces tend to activate a strongly pro-inflammatory innate immune response [79,94]. Particles with a negatively charged surface can inhibit the severity of the immune response by preventing antigen-presenting cells (APCs) from processing and presenting an antigen (biomaterial) for recognition by T cells [95].

Biomaterial surface topology provides a powerful tool to control and regulate corneal cell behavior [96], including cell adhesion [97], density, spreading, mobility [98], proliferation, differentiation [99], cytokine and ECM secretion [100,101], and cell signal transduction [102]. Importantly, the differentiation of keratocytes into myofibroblasts is triggered by the surface topography [103]. Thus, the surface topology of the biomaterial can inhibit the TGF-β-induced differentiation of myofibroblasts and prevent the development of fibrosis and corneal opacity during the healing process. Moreover, the differentiation of keratocytes into myofibroblasts is regulated by surface topography. Myrna et al. found that transformation into myofibroblasts could be prevented by cultured keratocytes on patterned grooves with a 1400-nm-wide pitch [103].

3.2.5. Anti-Oxidative Properties

Since extensive oxidative stress can occur in the implantation site, antioxidant properties of the biomaterial would be helpful. High-molecular-weight HA [104] and chitosan [105] have intrinsic anti-inflammatory properties due to their ROS-scavenging abilities.

3.2.6. Immune Cells

Activated neutrophils are recruited from the peripheral bloodstream by chemoattractant factors, adhere at the implantation site (via β2 integrins), and try to degrade the biomaterial by phagocytosis, proteolytic enzymes, and reactive oxygen species [79].

Increased immunomodulatory cytokines IL-10 and IL-17 are critical for corneal graft survival [74]. Treatment with T regulatory cells (Tregs) or tolerogenic APCs induced by immunoregulatory factors can help restore immune privilege and thus lead to the long-term survival of the corneal allograft in high-risk recipients. Host alloimmunity is the main cause of loss of donor CEnCs after corneal transplantation [106]. Tregs play a critical role in suppressing immune responses after tissue transplantation. Tregs from low-risk hosts can protect CEnCs from both Teff-mediated and IFN-γ- and TNF-α-induced cell death. This function is significantly compromised in Tregs derived from high-risk hosts. The cytoprotective role of Tregs is mediated by the immunomodulatory cytokine IL-10; hence, IL-10 is effective in protecting CEnC from inflammatory cytokines during cell death [106].

Keeping in mind all the data discussed above, it can be concluded that the number of molecular signaling pathways activated in the response to corneal trauma and biomaterial implantation play a pivotal role in the immunological response to the transplant. Modulating the activation/inhibition of involved molecular pathways along with proper biomaterial composition, surface topology and other parameters may provide a solution for establishing optimal host–graft interaction and ensure successful tissue regeneration, graft integration, and long-term survival.

4. Nanotechnology in Corneal Tissue Engineering

Nanotechnologies can be used at the stage of corneal scaffold fabrication to improve their physicochemical properties, but also after scaffold implantation, for example to deliver various therapeutic agents by means of nanocarriers in order to solve the problems of inflammation, secondary infections, and neovascularization in the damaged area.

4.1. Nanostructured Matrices

Nanoscaffolds possess unique mechanical properties that facilitate gas and nutrient exchange as well as the removal of cellular waste, and which also promote cell adhesion, proliferation, and differentiation [107]. For example, nanostructured ~10 nm dendrimers are high-contrast polymers that have a 3D ionic shape with numerous end groups. The biggest advantages of dendritic systems are the high density of functional side chains, the ability to manage network crosslinks, and scalability over a wide size range [108]. Dendrimer-based hydrogels have been shown to promote the rapid healing of corneal wounds without scarring or inflammation [109]. Due to the ability to control the crosslinking process and change the crosslinking chemistry, it is possible to manipulate the period of resorption, and thus control the process of wound healing on a bigger time scale. Thus, dendrimers are labile "smart" NMs and can be used for wound healing during long recovery periods with a low chance of an inflammatory response [110].

Another promising direction is the combined application of nanotechnology and corneal tissue engineering with natural biomaterials. For example, to form a biomaterial with the desired properties it can be combined with metal nanoparticles, graphene oxide, carbon nanotubes, and nanoliposomes [111]. Nanostructured hydrogels are mainly used for the delivery of genes and proteins. In situ transition from sol to gel promotes their role in enhancing the growth and functionality of other stem cells [112]. Soft nanoparticles can interact with polymer chains and can contribute to further crosslinking of the hydrogel grid to improve its mechanical properties [113].

4.2. Nanocarriers for Intracorneal Drug Delivery

Nanocarriers can improve the bioavailability and bio-distribution of therapeutic molecules, at the same time promoting targeted delivery and controlled drug release [114].

The problem of secondary infections after scaffold implantation can be addressed via the application of antibiotics, anti-viral, or anti-fungal drugs encapsulated within nanocarriers. For example, speaking about anti-viral drugs, dipeptide-acyclovir-based prodrugs encapsulated in poly (lactic-glycolic acid) (PLGA) nanoparticles showed increased efficacy due to improved drug release kinetics [115], while liposomes loaded with idoxuridine were reported to demonstrate increased penetration of the drug into the cornea [116]. There is also evidence that the retention time of the antifungal drug natamika delivered via chitosan nanoparticles in the corneal epithelial layer is 1.5 times longer than when using a commercial treatment method [117]. Speaking about antibiotics, quinolones moxifloxacin [118,119], sparfloxacin [120], and levofloxacin [121] demonstrated increased bioavailability when delivered via nanocarriers; better corneal permeability was reported as well. Sometimes nanoparticles can even be used as an alternative to antibiotics, for example, as silver ones can, known for their remarkable antiseptic properties [122].

Another important issue in corneal tissue engineering is inflammation and subsequent neovascularization, which threatens corneal transparency. Currently available options to avoid this condition include corticosteroids, non-steroidal anti-inflammatory eye drops, photodynamic therapy, photocoagulation, and antibodies (bevacizumab) against vascular endothelial growth factor A (VEGF-A). These methods aim to suppress angiogenesis by blocking angiogenic factors such as VEGF, PDGF, major fibroblast growth factor (FGF), MMPs, and interleukins [123–125]. For example, in a study conducted by Iriyama et al., micelles consisting of a copolymer and plasmid DNA expressing the soluble VEGF receptor 1 (sFit-1) were used for gene therapy [126]. SFit-1 expression acted as a VEGF receiver and prevented activation of the angiogenesis cascade. The results showed that the injection of micelles containing a reporter gene lead to delayed sFit-1 expression and inhibition of corneal neovascularization. Gold nanoparticles were also reported to inhibit angiogenesis [127] and corneal neovascularization [128] by suppressing the vascular expression of endothelial growth factor (EGF) receptor-2.

Dexamethasone, a widely used anti-inflammatory steroid drug, was reported to show higher bioavailability and better corneal penetration when delivered via nanomicelles [129] or encapsulated within nanoparticles [118,130], while hydrocortisone, another widely used steroid drug, was reported to demonstrate fewer dose-dependent side effects when administered in the form of nanosuspension [131].

Rapid re-epithelialization is one of the key factors preventing infection and promoting implant/host integration. Xuan Zhao et al. used the complexes of AuNPs and microRNA-133b sorbed on collagen matrices to restore the cornea and inhibit scarring [132], reporting good and fast re-epithelialization.

5. Outlook and Future Perspectives

Regenerative ophthalmology is a rapidly evolving new field for the regeneration of lost or damaged eye cells and tissues as well as the treatment of vision loss and blindness caused by various ocular diseases, injuries, or infections. However, the cell therapy approaches in regenerative medicine are still at an early stage of development and face numerous serious problems and challenges. Effective methods and biomaterials for transplantation should support the correct rate of cell adhesion, proliferation, and differentiation, and sustain the desired cellular phenotype, cell-specific signaling, and biochemical properties. The use of combination therapy of nanomedicine/bioengineering in ocular regeneration is promising in overcoming these difficulties [133]. The possibility to introduce controlled and customizable changes in the psychochemical characteristics, size, and surface chemistry of NMs allows the construction of various matrixes with desired properties, perfectly optimized for specific biological applications (Figure 1). Non-viral gene nanocarriers such as polyplexes, mesoporous NPs, organic–inorganic hybrid nanocarriers, nanoscripts, self-organizing DNA nanostructures, and magnetic NPs are becoming promising tools for reprogramming cells on the way to treat and regenerate damaged corneal tissues. The combination of nanotechnology and immunoengineering to modulate the innate and adaptive immune response will be crucial for corneal wound healing.

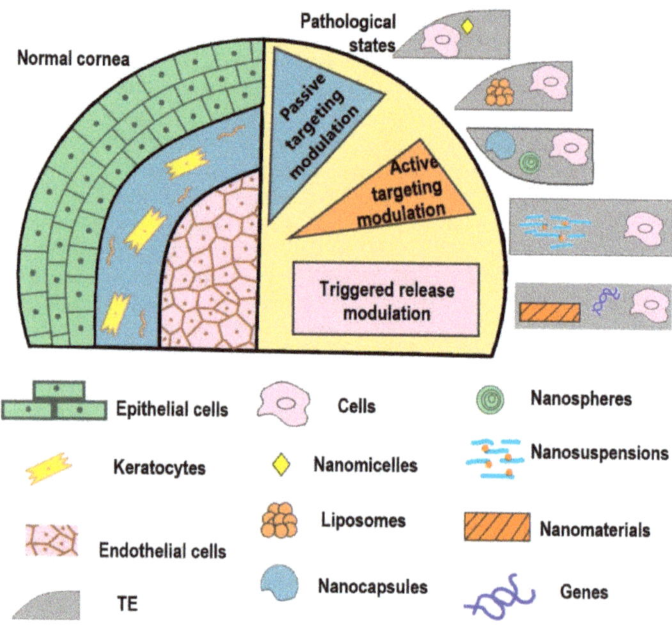

Figure 1. Nanomaterial-based approach in the combined therapy. Top left – healthy corneal tissue. The cornea is a complex vascular-free structure composed of five layers, three of which are of cellular nature (epithelium, stroma and endothelium). Throughout the lifetime, corneal cells are exposed to various traumatic and damaging factors from the external environment and inner disturbances in organism's functionality. These pathological processes and damaging agents can compromise the integrity of the cornea and lead to vision loss. Top right – pathological corneal tissue. Due to the lack of modern approaches that allow the full restoration of the cornea tissue and vision, new treatment and therapeutic strategies are needed to be introduced. TE- and nanotechnology-based strategies can become a new chapter in the cornea restoration. TE constructs can act through active and passive targeting and controlled triggered release promoting the most effective approach for each set of specific molecular mechanisms and cellular events.

Personalized medicine that adapts the treatment of a disease based on an individual's genetics and specific pathophysiological processes taking place in their specific case to achieve a better clinical outcome is an emerging and growing field in ophthalmology. The use of nanotechnology can allow personalized therapy and optimized drug dosages for more accurate and effective treatment of corneal diseases. It is expected that in the future nanotechnology will be used to personalize regenerative medicine using human stem cells and provide therapeutic tools to maintain a healthy environment for the growth and maturation of stem cells in the damaged area [134]. However, research in nanotechnology for regenerative ophthalmology is still at an early stage, and there is a very limited number of in vivo studies. The behavior of corneal cells on TE constructs in the area of corneal damage has been widely demonstrated in vitro, but many open questions remain due to the lack of in vivo proof-of-concept studies.

Author Contributions: O.M. drafted and wrote the manuscript, T.P. contributed to the preparation of "Nanotechnology in Corneal Tissue Engineering " section, A.N. contributed to the preparation of " Corneal Tissue Engineering " section, M.A. and A.B. contributed to the preparation of " Molecular Pathways and Interactions between Host Tissues and a Graft " section, M.P. and P.B. prepared the manuscript revision, B.T. and S.B. contributed to the design of the study, S.A. provided critical revision of the manuscript, and P.T. contributed to the design of the study and provided critical revision of the manuscript. All authors have read and agreed to the published version of the manuscript.

Funding: This research was funded by the Ministry of Science and Higher Education of the Russian Federation within the framework of state support for the creation and development of World-class research center "Digital biodesign and personalized healthcare", grant number NO. 075-15-2020-926.

Conflicts of Interest: The authors declare no conflict of interest.

Abbreviations

3D	Three-dimensional
ABCG2	ATP-binding cassette super-family G member 2
AM	Amniotic membrane
BMP4	Bone morphogenetic protein 4
CECs	Corneal epithelial cells
CenCs	Corneal endothelial cells
DC	Decellularized cornea
ECM	Extracellular matrix
ER	Endoplasmic reticulum
FAK	Focal adhesion kinase
hCECs	Human CECs
LESCs	Limbal epithelial stem cells
LSCs	Limbal stem cells
MMP	Matrix metallopeptidases
MSCs	Mesenchymal stem cells
NM	Nanomaterials
NPs	Nanoparticles
PCL	Polycaprolactone
PEG	Polyethylene glycol
PMMA	Poly (methyl methacrylate)
PLGA	Poly lacto-glycolic acid
RF	Riboflavin
RhoA	Ras homolog family member A
sFit-1	Soluble VEGF receptor 1
TE	Tissue-engineered
Tregs	T regulatory cells
VEGF-A	Vascular endothelial growth factor A
YAP	Yes-associated protein
α-SMA	Smooth muscle alpha actin (alpha smooth muscle actin)

References

1. Morishige, N.; Shin-Gyou-Uchi, R.; Azumi, H.; Ohta, H.; Morita, Y.; Yamada, N.; Kimura, K.; Takahara, A.; Sonoda, K.H. Quantitative analysis of collagen lamellae in the normal and keratoconic human cornea by second harmonic generation imaging microscopy, Investig. Ophthalmol. *Vis. Sci.* **2014**, *55*, 8377–8385. [CrossRef]
2. Mobaraki, M.; Abbasi, R.; Vandchali, S.O.; Ghaffari, M.; Moztarzadeh, F.; Mozafari, M. Corneal repair and regeneration: Current concepts and future directions. *Front. Bioeng. Biotechnol.* **2019**, *7*, 135. [CrossRef]
3. Gonzalez, G.; Sasamoto, Y.; Ksander, B.R.; Frank, M.H.; Frank, N.Y. Limbal Stem Cells: Identity, Developmental Origin and Therapeutic Potential, Wiley Interdiscip. *Rev. Dev. Biol.* **2018**, *7*. [CrossRef]
4. Gain, P.; Jullienne, R.; He, Z.; Aldossary, M.; Acquart, S.; Cognasse, F.; Thuret, G. Global survey of corneal transplantation and eye banking. *JAMA Ophthalmol.* **2016**, *134*, 167–173. [CrossRef] [PubMed]
5. Hamrah, Y.Q.P. Corneal Allograft Rejection: Immunopathogenesis to Therapeutics. *J. Clin. Cell. Immunol.* **2013**. [CrossRef]
6. Matthyssen, S.; van den Bogerd, B.; Dhubhghaill, S.N.; Koppen, C.; Zakaria, N. Corneal regeneration: A review of stromal replacements. *Acta Biomater.* **2018**, *69*, 31–41. [CrossRef]

7. Susaimanickam, P.J.; Maddileti, S.; Pulimamidi, V.K.; Boyinpally, S.R.; Naik, R.R.; Naik, M.N.; Reddy, G.B.; Sangwan, V.S.; Mariappan, I. Generating minicorneal organoids from human induced pluripotent stem cells. *Development* **2017**, *144*, 2338–2351. [CrossRef]
8. Mahdavi, S.S.; Abdekhodaie, M.J.; Mashayekhan, S.; Baradaran-Rafii, A.; Djalilian, A.R. Bioengineering Approaches for Corneal Regenerative Medicine. *Tissue Eng. Regen. Med.* **2020**. [CrossRef]
9. Ljubimov, A.V.; Saghizadeh, M. Progress in corneal wound healing, Prog. *Retin. Eye Res.* **2015**, *49*, 17–45.
10. Zarei-Ghanavati, M.; Liu, C. Aspects of corneal transplant immunology. *J. Ophthalmic Vis. Res.* **2017**. [CrossRef] [PubMed]
11. Amouzegar, A.; Chauhan, S.K.; Dana, R. Alloimmunity and Tolerance in Corneal Transplantation. *J. Immunol.* **2016**. [CrossRef]
12. Hori, J.; Yamaguchi, T.; Keino, H.; Hamrah, P.; Maruyama, K. Immune privilege in corneal transplantation. *Prog. Retin. Eye Res.* **2019**, *72*. [CrossRef]
13. Fuchsluger, T.A.; Jurkunas, U.; Kazlauskas, A.; Dana, R. Corneal endothelial cells are protected from apoptosis by gene therapy. *Hum. Gene Ther.* **2011**, *22*, 549–558. [CrossRef] [PubMed]
14. Vicente-Pascual, M.; Albano, A.; Solinís, M.; Serpe, L.; Rodríguez-Gascón, A.; Foglietta, F.; Muntoni, E.; Torrecilla, J.; del Pozo-Rodríguez, A.; Battaglia, L. Gene delivery in the cornea: In vitro & ex vivo evaluation of solid lipid nanoparticle-based vectors. *Nanomedicine* **2018**, *13*, 1847–1864. [CrossRef] [PubMed]
15. Raub, C.B.; Orwin, E.J.; Haskell, R. Immunogold labeling to enhance contrast in optical coherence microscopy of tissue engineered corneal constructs. In *Annual International Conference IEEE Engineering Medicine Biology*; IEEE: Piscataway, NJ, USA, 2004; pp. 1210–1213. [CrossRef]
16. Lee, C.-J.; Buznyk, O.; Kuffova, L.; Rajendran, V.; Forrester, J.V.; Phopase, J.; Islam, M.M.; Skog, M.; Ahlqvist, J.; Griffith, M. Cathelicidin LL-37 and HSV-1 Corneal Infection: Peptide Versus Gene Therapy. *Transl. Vis. Sci. Technol.* **2014**, *3*, 4. [CrossRef] [PubMed]
17. Chomchalao, P.; Nimtrakul, P.; Pham, D.T.; Tiyaboonchai, W. Development of amphotericin B-loaded fibroin nanoparticles: A novel approach for topical ocular application. *J. Mater. Sci.* **2020**, *55*, 5268–5279. [CrossRef]
18. Bektas, C.K.; Hasirci, V. Cell Loaded GelMA:HEMA IPN hydrogels for corneal stroma engineering. *J. Mater. Sci. Mater. Med.* **2019**, *31*, 1–15. [CrossRef]
19. Shavkuta, B.S.; Gerasimov, M.Y.; Minaev, N.V.; Kuznetsova, D.S.; Dudenkova, V.V.; Mushkova, I.A.; Malyugin, B.E.; Kotova, S.L.; Timashev, P.S.; Kostenev, S.V.; et al. Highly effective 525 nm femtosecond laser crosslinking of collagen and strengthening of a human donor cornea. *Laser Phys. Lett.* **2018**, *15*, 015602. [CrossRef]
20. Jo, D.H.; Lee, T.G.; Kim, J.H. Nanotechnology and nanotoxicology in retinopathy. *Int. J. Mol. Sci.* **2011**, *12*, 8288–8301. [CrossRef]
21. Du, L.; Wu, X. Development and Characterization of a Full-Thickness Acellular Porcine Cornea Matrix for Tissue Engineering. *Artif. Organs.* **2011**, *35*, 691–705. [CrossRef] [PubMed]
22. Wu, Z.; Zhou, Y.; Li, N.; Huang, M.; Duan, H.; Ge, J.; Xiang, P.; Wang, Z. The use of phospholipase A2 to prepare acellular porcine corneal stroma as a tissue engineering scaffold. *Biomaterials* **2009**, *30*, 3513–3522. [CrossRef] [PubMed]
23. Palchesko, R.N.; Carrasquilla, S.D.; Feinberg, A.W. Natural Biomaterials for Corneal Tissue Engineering, Repair, and Regeneration. *Adv. Healthc. Mater.* **2018**, *7*, 1701434. [CrossRef] [PubMed]
24. Fernández-Pérez, J.; Ahearne, M. The impact of decellularization methods on extracellular matrix derived hydrogels. *Sci. Rep.* **2019**, *9*. [CrossRef] [PubMed]
25. Chakraborty, J.; Roy, S.; Murab, S.; Ravani, R.; Kaur, K.; Devi, S.; Singh, D.; Sharma, S.; Mohanty, S.; Dinda, A.K.; et al. Modulation of Macrophage Phenotype, Maturation, and Graft Integration through Chondroitin Sulfate Cross-Linking to Decellularized Cornea. *ACS Biomater. Sci. Eng.* **2019**, *5*, 165–179. [CrossRef] [PubMed]
26. Wilson, S.; Sidney, L.; Dunphy, S.; Rose, J.; Hopkinson, A. Keeping an Eye on Decellularized Corneas: A Review of Methods, Characterization and Applications. *J. Funct. Biomater.* **2013**, *4*, 114–161. [CrossRef] [PubMed]
27. Porzionato, A.; Stocco, E.; Barbon, S.; Grandi, F.; Macchi, V.; de Caro, R. Tissue-engineered grafts from human decellularized extracellular matrices: A systematic review and future perspectives. *Int. J. Mol. Sci.* **2018**, *19*, 4117. [CrossRef] [PubMed]
28. Scalcione, C.; Ortiz-Vaquerizas, D.; Said, D.G.; Dua, H.S. Fibrin glue as agent for sealing corneal and conjunctival wound leaks. *Eye* **2018**, *32*, 463–466. [CrossRef]
29. Jacob, S.; Dhawan, P.; Tsatsos, M.; Agarwal, A.; Narasimhan, S.; Kumar, A. Fibrin Glue–Assisted Closure of Macroperforation in Predescemetic Deep Anterior Lamellar Keratoplasty With a Donor Obtained From Small Incision Lenticule Extraction. *Cornea* **2019**, *38*, 775–779. [CrossRef] [PubMed]
30. Rizwan, M.; Peh, G.S.L.; Ang, H.P.; Lwin, N.C.; Adnan, K.; Mehta, J.S.; Tan, W.S.; Yim, E.K.F. Sequentially-crosslinked bioactive hydrogels as nano-patterned substrates with customizable stiffness and degradation for corneal tissue engineering applications. *Biomaterials* **2017**, *120*, 139–154. [CrossRef]
31. Crawford, G.J. The development and results of an artificial cornea: AlphaCor^{TM}. *Biomater. Regen. Med. Ophthalmol.* **2016**, 443–462. [CrossRef]
32. Nouri, M.; Terada, H.; Alfonso, E.C.; Foster, C.S.; Durand, M.L.; Dohlman, C.H. Endophthalmitis after keratoprosthesis: Incidence, bacterial causes, and risk factors. *Arch. Ophthalmol.* **2001**, *119*, 484–489. [CrossRef]
33. Rico-Sánchez, L.; Garzón, I.; González-Andrades, M.; Ruíz-García, A.; Punzano, M.; Lizana-Moreno, A.; Muñoz-Ávila, J.I.; Sánchez-Quevedo, M.d.; Martínez-Atienza, J.; Lopez-Navas, L.; et al. Successful development and clinical translation of a novel anterior lamellar artificial cornea. *J. Tissue Eng. Regen. Med.* **2019**, *13*, 2142–2154. [CrossRef]

34. Yeung, A.M.; Faraj, L.A.; McIntosh, O.D.; Dhillon, V.K.; Dua, H.S. Fibrin glue inhibits migration of ocular surface epithelial cells. *Eye* **2016**, *30*, 1389–1394. [CrossRef]
35. Wang, J.; Chen, Y.; Bai, Y.; Quan, D.; Wang, Z.; Xiong, L.; Shao, Z.; Sun, W.; Mi, S. A core-skirt designed artificial cornea with orthogonal microfiber grid scaffold. *Exp. Eye Res.* **2020**, *195*, 108037. [CrossRef] [PubMed]
36. Ma, D.H.K.; Chen, H.C.; Ma, K.S.K.; Lai, J.Y.; Yang, U.; Yeh, L.K.; Hsueh, Y.J.; Chu, W.K.; Lai, C.H.; Chen, J.K. Preservation of human limbal epithelial progenitor cells on carbodiimide cross-linked amniotic membrane via integrin-linked kinase-mediated Wnt activation. *Acta Biomater.* **2016**, *31*, 144–155. [CrossRef] [PubMed]
37. Zhou, Z.; Long, D.; Hsu, C.C.; Liu, H.; Chen, L.; Slavin, B.; Lin, H.; Li, X.; Tang, J.; Yiu, S.; et al. Nanofiber-reinforced decellularized amniotic membrane improves limbal stem cell transplantation in a rabbit model of corneal epithelial defect. *Acta Biomater.* **2019**, *97*, 310–320. [CrossRef]
38. Haj, E.; Sidney, L.E.; Patient, J.; White, L.J.; Dua, H.S.; El Haj, A.J.; Hopkinson, A.; Rose, F.R. In vitro evaluation of electrospun blends of gelatin and PCL for application as a partial thickness corneal graft. *J. Biomed. Mater. Res. Part A.* **2018**, *107*. [CrossRef]
39. Wang, Y.H.; Young, T.H.; Wang, T.J. Investigating the effect of chitosan/ polycaprolactone blends in differentiation of corneal endothelial cells and extracellular matrix compositions. *Exp. Eye Res.* **2019**, *185*, 107679. [CrossRef] [PubMed]
40. Song, J.E.; Sim, B.R.; Jeon, Y.S.; Kim, H.S.; Shin, E.Y.; Carlomagno, C.; Khang, G. Characterization of surface modified glycerol/silk fibroin film for application to corneal endothelial cell regeneration. *J. Biomater. Sci. Polym. Ed.* **2019**, *30*, 263–275. [CrossRef] [PubMed]
41. Yamaguchi, H.; Takezawa, T. Fabrication of a corneal model composed of corneal epithelial and endothelial cells via a collagen vitrigel membrane functioned as an acellular stroma and its application to the corneal permeability test of chemicals. *Drug Metab. Dispos.* **2018**, *46*, 1684–1691. [CrossRef] [PubMed]
42. Yoshida, J.; Yokoo, S.; Oshikata-Miyazaki, A.; Amano, S.; Takezawa, T.; Yamagami, S. Transplantation of Human Corneal Endothelial Cells Cultured on Bio-Engineered Collagen Vitrigel in a Rabbit Model of Corneal Endothelial Dysfunction. *Curr. Eye Res.* **2017**, *42*, 1420–1425. [CrossRef] [PubMed]
43. Shah, R.; Stodulka, P.; Skopalova, K.; Saha, P. Dual Crosslinked Collagen/Chitosan Film for Potential Biomedical Applications. *Polymers* **2019**, *11*, 2094. [CrossRef]
44. Hatami-Marbini, H.; Jayaram, S.M. UVA/riboflavin collagen crosslinking stiffening effects on anterior and posterior corneal flaps. *Exp. Eye Res.* **2018**, *176*, 53–58. [CrossRef]
45. Long, Y.; Zhao, X.; Liu, S.; Chen, M.; Liu, B.; Ge, J.; Jia, Y.G.; Ren, L. Collagen–hydroxypropyl methylcellulose membranes for corneal regeneration. *ACS Omega* **2018**, *3*, 1269–1275. [CrossRef] [PubMed]
46. In situ-forming bio-orthogonally crosslinked collagen-hyaluronate co-polymeric hydrogel to treat deep corneal stromal defects: In vivo biological response. *IOVS ARVO J.* **2020**, *61*, 1208.
47. Rose, J.B.; Pacelli, S.; el Haj, A.J.; Dua, H.S.; Hopkinson, A.; White, L.J.; Rose, F.R.A.J. Gelatin-based materials in ocular tissue engineering. *Materials* **2014**, *7*, 3106–3135. [CrossRef] [PubMed]
48. Goodarzi, H.; Jadidi, K.; Pourmotabed, S.; Sharifi, E.; Aghamollaei, H. Preparation and in vitro characterization of cross-linked collagen–gelatin hydrogel using EDC/NHS for corneal tissue engineering applications. *Int. J. Biol. Macromol.* **2019**, *126*, 620–632. [CrossRef] [PubMed]
49. Xu, W.; Liu, K.; Li, T.; Zhang, W.; Dong, Y.; Lv, J.; Wang, W.; Sun, J.; Li, M.; Wang, M.; et al. An in situ hydrogel based on carboxymethyl chitosan and sodium alginate dialdehyde for corneal wound healing after alkali burn. *J. Biomed. Mater. Res. Part A* **2018**, *107*, 36589. [CrossRef]
50. Stafiej, P.; Schubert, D.W.; Fuchsluger, T. Mechanical and Optical Properties of PCL Nanofiber Reinforced Alginate Hydrogels for Application in Corneal Wound Healing Draw resonance View project biomaterials for corneal tissue engineering and wound healing View project. *Biomater. Med. Appl.* **2018**. [CrossRef]
51. Tsai, C.Y.; Woung, L.C.; Yen, J.C.; Tseng, P.C.; Chiou, S.H.; Sung, Y.J.; Liu, K.T.; Cheng, Y.H. Thermosensitive chitosan-based hydrogels for sustained release of ferulic acid on corneal wound healing. *Carbohydr. Polym.* **2016**, *135*, 308–315. [CrossRef]
52. Tang, Q.; Luo, C.; Lu, B.; Fu, Q.; Yin, H.; Qin, Z.; Lyu, D.; Zhang, L.; Fang, Z.; Zhu, Y.; et al. Thermosensitive chitosan-based hydrogels releasing stromal cell derived factor-1 alpha recruit MSC for corneal epithelium regeneration. *Acta Biomater.* **2017**, *61*, 101–113. [CrossRef] [PubMed]
53. Suárez-Barrio, C.; Etxebarria, J.; Hernáez-Moya, R.; del Val-Alonso, M.; Rodriguez-Astigarraga, M.; Urkaregi, A.; Freire, V.; Morales, M.-C.; Durán, J.; Vicario, M.; et al. Hyaluronic Acid Combined with Serum Rich in Growth Factors in Corneal Epithelial Defects. *Int. J. Mol. Sci.* **2019**, *20*, 1655. [CrossRef] [PubMed]
54. Zhong, J.; Deng, Y.; Tian, B.; Wang, B.; Sun, Y.; Huang, H.; Chen, L.; Ling, S.; Yuan, J. Hyaluronate acid-dependent protection and enhanced corneal wound healing against oxidative damage in corneal epithelial cells. *J. Ophthalmol.* **2016**, *2016*. [CrossRef]
55. Biazar, E.; Baradaran-Rafii, A.; Heidari-Keshel, S.; Tavakolifard, S. Oriented nanofibrous silk as a natural scaffold for ocular epithelial regeneration. *J. Biomater. Sci. Polym. Ed.* **2015**, *26*, 1139–1151. [CrossRef] [PubMed]
56. Aghaei-Ghareh-Bolagh, B.; Guan, J.; Wang, Y.; Martin, A.D.; Dawson, R.; Mithieux, S.M.; Weiss, A.S. Optically robust, highly permeable and elastic protein films that support dual cornea cell types. *Biomaterials* **2019**, *188*, 50–62. [CrossRef]
57. Xu, W.; Wang, Z.; Liu, Y.; Wang, L.; Jiang, Z.; Li, T.; Zhang, W.; Liang, Y. Carboxymethyl chitosan/gelatin/hyaluronic acid blended-membranes as epithelia transplanting scaffold for corneal wound healing. *Carbohydr. Polym.* **2018**, *192*, 240–250. [CrossRef] [PubMed]

58. Sabater, A.L.; Perez, V.L. Amniotic membrane use for management of corneal limbal stem cell deficiency. *Curr. Opin. Ophthalmol.* **2017**, *28*, 363–369. [CrossRef] [PubMed]
59. Mingo-Botín, D.; Balgos, M.J.T.D.; Arnalich-Montiel, F. *Corneal Endothelium: Isolation and Cultivation Methods*; Springer: Cham, Switzerland, 2019; pp. 425–436. [CrossRef]
60. Che, X.; Wu, H.; Jia, C.; Sun, H.; Ou, S.; Wang, J.; Jeyalatha, M.V.; He, X.; Yu, J.; Zuo, C.; et al. A novel tissue-engineered corneal stromal equivalent based on amniotic membrane and keratocytes. *Investig. Ophthalmol. Vis. Sci.* **2019**, *60*, 517–527. [CrossRef] [PubMed]
61. Zhang, L.; Zou, D.; Li, S.; Wang, J.; Qu, Y.; Ou, S.; Jia, C.; Li, J.; He, H.; Liu, T.; et al. An Ultra-thin Amniotic Membrane as Carrier in Corneal Epithelium Tissue-Engineering. *Sci. Rep.* **2016**, *6*, 1–12. [CrossRef] [PubMed]
62. Koh, H.; Sirivisoot, S.; Talchai, S.; Khemarangsan, V. Efficient mesenchymal stem cell conversion to corneal keratocytes by the use of specific biomaterial and Cell Rev^{TM} mSC diffKera media. *Cytotherapy* **2020**, *22*, S182. [CrossRef]
63. Gouveia, R.M.; Lepert, G.; Gupta, S.; Mohan, R.R.; Paterson, C.; Connon, C.J. Assessment of corneal substrate biomechanics and its effect on epithelial stem cell maintenance and differentiation. *Nat. Commun.* **2019**, *10*. [CrossRef] [PubMed]
64. Fernández-Pérez, J.; Kador, K.E.; Lynch, A.P.; Ahearne, M. Characterization of extracellular matrix modified poly(ε-caprolactone) electrospun scaffolds with differing fiber orientations for corneal stroma regeneration. *Mater. Sci. Eng. C* **2020**, *108*, 110415. [CrossRef] [PubMed]
65. Ludwig, P.; Huff, T.; Zuniga, J. The potential role of bioengineering and three-dimensional printing in curing global corneal blindness. *J. Tissue Eng.* **2018**. [CrossRef]
66. Anderson, J.M.; Rodriguez, A.; Chang, D.T. Foreign body reaction to biomaterials. *Semin. Immunol.* **2008**. [CrossRef]
67. Crupi, A.; Costa, A.; Tarnok, A.; Melzer, S.; Teodori, L. Inflammation in tissue engineering: The Janus between engraftment and rejection. *Eur. J. Immunol.* **2015**. [CrossRef]
68. Kobayashi, T.; Shiraishi, A.; Hara, Y.; Kadota, Y.; Yang, L.; Inoue, T.; Shirakata, Y.; Ohashi, Y. Stromal-epithelial interaction study: The effect of corneal epithelial cells on growth factor expression in stromal cells using organotypic culture model. *Exp. Eye Res.* **2015**, *135*, 109–117. [CrossRef]
69. West-Mays, J.A.; Dwivedi, D.J. The keratocyte: Corneal stromal cell with variable repair phenotypes. *Int. J. Biochem. Cell Biol.* **2006**. [CrossRef] [PubMed]
70. Torricelli, A.A.M.; Santhanam, A.; Wu, J.; Singh, V.; Wilson, S.E. The corneal fibrosis response to epithelial-Stromal injury. *Exp. Eye Res.* **2016**, *142*, 110–118. [CrossRef]
71. Guo, Y.; Ma, X.; Wu, W.; Shi, M.; Ma, J.; Zhang, Y.; Zhao, E.; Yang, X. Coordinated microRNA/mRNA expression profiles reveal a putative mechanism of corneal epithelial cell transdifferentiation from skin epidermal stem cells. *Int. J. Mol. Med.* **2018**, *41*, 877–887. [CrossRef]
72. Mitra, S.K.; Schlaepfer, D.D. Integrin-regulated FAK-Src signaling in normal and cancer cells. *Curr. Opin. Cell Biol.* **2006**, *18*, 516–523. [CrossRef] [PubMed]
73. Osaki, T.; Serrano, J.C.; Kamm, R.D. Cooperative Effects of Vascular Angiogenesis and Lymphangiogenesis. *Regen. Eng. Transl. Med.* **2018**, *4*, 120–132. [CrossRef]
74. Couture, C.; Zaniolo, K.; Carrier, P.; Lake, J.; Patenaude, J.; Germain, L.; Guérin, S.L. The tissue-engineered human cornea as a model to study expression of matrix metalloproteinases during corneal wound healing. *Biomaterials* **2016**, *78*, 86–101. [CrossRef] [PubMed]
75. Omoto, M.; Suri, K.; Amouzegar, A.; Li, M.; Katikireddy, K.R.; Mittal, S.K.; Chauhan, S.K. Hepatocyte Growth Factor Suppresses Inflammation and Promotes Epithelium Repair in Corneal Injury. *Mol. Ther.* **2017**, *25*, 1881–1888. [CrossRef]
76. Salabarria, A.; Braun, G.; Heykants, M.; Koch, M.; Reuten, R.; Mahabir, E.; Cursiefen, C.; Bock, F. Local VEGF -A blockade modulates the microenvironment of the corneal graft bed. *Am. J. Transplant.* **2019**, *19*, 2446–2456. [CrossRef] [PubMed]
77. Han, Y.; Li, C.; Cai, Q.; Bao, X.; Tang, L.; Ao, H.; Liu, J.; Jin, M.; Zhou, Y.; Wan, Y.; et al. Studies on bacterial cellulose/poly(vinyl alcohol) hydrogel composites as tissue-engineered corneal stroma. *Biomed. Mater.* **2020**, *15*, 035022. [CrossRef] [PubMed]
78. Chen, Z.; You, J.; Liu, X.; Cooper, S.; Hodge, C.; Sutton, G.; Crook, J.M.; Wallace, G.G. Biomaterials for corneal bioengineering. *Biomed. Mater.* **2018**, *13*, 032002. [CrossRef] [PubMed]
79. Mariani, E.; Lisignoli, G.; Borzì, R.M.; Pulsatelli, L. Biomaterials: Foreign bodies or tuners for the immune response? *Int. J. Mol. Sci.* **2019**, 636. [CrossRef]
80. McNally, A.K.; Jones, J.A.; MacEwan, S.R.; Colton, E.; Anderson, J.M. Vitronectin is a critical protein adhesion substrate for IL-4-induced foreign body giant cell formation. *J. Biomed. Mater. Res. Part A* **2008**. [CrossRef] [PubMed]
81. Jenney, C.R.; Anderson, J.M. Adsorbed serum proteins responsible for surface dependent human macrophage behavior. *J. Biomed. Mater. Res.* **2000**. [CrossRef]
82. Keselowsky, B.G.; Bridges, A.W.; Burns, K.L.; Tate, C.C.; Babensee, J.E.; LaPlaca, M.C.; García, A.J. Role of plasma fibronectin in the foreign body response to biomaterials. *Biomaterials* **2007**. [CrossRef] [PubMed]
83. Sheikh, Z.; Brooks, P.J.; Barzilay, O.; Fine, N.; Glogauer, M. Macrophages, foreign body giant cells and their response to implantable biomaterials. *Materials* **2015**. [CrossRef]
84. McKay, T.B.; Hutcheon, A.E.K.; Guo, X.; Zieske, J.D.; Karamichos, D. Modeling the cornea in 3-dimensions: Current and future perspectives. *Exp. Eye Res.* **2020**, *197*, 108127. [CrossRef] [PubMed]

85. RGouveia, M.; Vajda, F.; Wibowo, J.A.; Figueiredo, F.; Connon, C.J. YAP, ΔNp63, and β-Catenin Signaling Pathways Are Involved in the Modulation of Corneal Epithelial Stem Cell Phenotype Induced by Substrate Stiffness. *Cells* **2019**, *8*, 347. [CrossRef]
86. Kim, Y.K.; Que, R.; Wang, S.W.; Liu, W.F. Modification of Biomaterials with a Self-Protein Inhibits the Macrophage Response. *Adv. Healthc. Mater.* **2014**. [CrossRef]
87. Seong, S.Y.; Matzinger, P. Hydrophobicity: An ancient damage-associated molecular pattern that initiates innate immune responses. *Nat. Rev. Immunol.* **2004**. [CrossRef]
88. Moyano, D.F.; Goldsmith, M.; Solfiell, D.J.; Landesman-Milo, D.; Miranda, O.R.; Peer, D.; Rotello, V.M. Nanoparticle hydrophobicity dictates immune response. *J. Am. Chem. Soc.* **2012**. [CrossRef] [PubMed]
89. Song, S.; Ravensbergen, K.; Alabanza, A.; Soldin, D.; Hahm, J.I. Distinct adsorption configurations and self-Assembly characteristics of fibrinogen on chemically uniform and alternating surfaces including block copolymer nanodomains. *ACS Nano* **2014**. [CrossRef] [PubMed]
90. Kumar, N.; Parajuli, O.; Gupta, A.; Hahm, J.I. Elucidation of protein adsorption behavior on polymeric surfaces: Toward high-density, high-payload protein templates. *Langmuir* **2008**. [CrossRef] [PubMed]
91. Ouberai, M.M.; Xu, K.; Welland, M.E. Effect of the interplay between protein and surface on the properties of adsorbed protein layers. *Biomaterials* **2014**. [CrossRef]
92. Mitrousis, N.; Fokina, A.; Shoichet, M.S. Biomaterials for cell transplantation. *Nat. Rev. Mater.* **2018**. [CrossRef]
93. Blaszykowski, C.; Sheikh, S.; Thompson, M. Biocompatibility and antifouling: Is there really a linkα. *Trends Biotechnol.* **2014**. [CrossRef]
94. Bartneck, M.; Keul, H.A.; Singh, S.; Czaja, K.; Bornemann, J.; Bockstaller, M.; Moeller, M.; Zwadlo-Klarwasser, G.; Groll, J. Rapid uptake of gold nanorods by primary human blood phagocytes and immunomodulatory effects of surface chemistry. *ACS Nano* **2010**. [CrossRef]
95. Wen, Y.; Waltman, A.; Han, H.; Collier, J.H. Switching the Immunogenicity of Peptide Assemblies Using Surface Properties. *ACS Nano* **2016**. [CrossRef] [PubMed]
96. Christo, S.N.; Bachhuka, A.; Diener, K.R.; Mierczynska, A.; Hayball, J.D.; Vasilev, K. The Role of Surface Nanotopography and Chemistry on Primary Neutrophil and Macrophage Cellular Responses. *Adv. Healthc. Mater.* **2016**. [CrossRef] [PubMed]
97. Hulander, M.; Lundgren, A.; Faxälv, L.; Lindahl, T.L.; Palmquist, A.; Berglin, M.; Elwing, H. Gradients in surface nanotopography used to study platelet adhesion and activation. *Colloids Surfaces B Biointerfaces* **2013**. [CrossRef]
98. Gilchrist, C.L.; Ruch, D.S.; Little, D.; Guilak, F. Micro-scale and meso-scale architectural cues cooperate and compete to direct aligned tissue formation. *Biomaterials* **2014**. [CrossRef] [PubMed]
99. Gittens, R.A.; McLachlan, T.; Olivares-Navarrete, R.; Cai, Y.; Berner, S.; Tannenbaum, R.; Schwartz, Z.; Sandhage, K.H.; Boyan, B.D. The effects of combined micron-/submicron-scale surface roughness and nanoscale features on cell proliferation and differentiation. *Biomaterials* **2011**. [CrossRef]
100. Ma, Q.L.; Zhao, L.Z.; Liu, R.R.; Jin, B.Q.; Song, W.; Wang, Y.; Zhang, Y.S.; Chen, L.H.; Zhang, Y.M. Improved implant osseointegration of a nanostructured titanium surface via mediation of macrophage polarization. *Biomaterials* **2014**. [CrossRef]
101. Tan, K.S.; Qian, L.; Rosado, R.; Flood, P.M.; Cooper, L.F. The role of titanium surface topography on J774A.1 macrophage inflammatory cytokines and nitric oxide production. *Biomaterials* **2006**. [CrossRef] [PubMed]
102. Xiong, S.; Gao, H.C.; Qin, L.; Jia, Y.G.; Ren, L. Engineering topography: Effects on corneal cell behavior and integration into corneal tissue engineering. *Bioact. Mater.* **2019**, *4*, 293–302. [CrossRef]
103. Myrna, K.E.; Mendonsa, R.; Russell, P.; Pot, S.A.; Liliensiek, S.J.; Jester, J.V.; Nealey, P.F.; Brown, D.; Murphy, C.J. Substratum topography modulates corneal fibroblast to Myofibroblast transformation. *Investig. Ophthalmol. Vis. Sci.* **2012**. [CrossRef]
104. Nakamura, K.; Yokohama, S.; Yoneda, M.; Okamoto, S.; Tamaki, Y.; Ito, T.; Okada, M.; Aso, K.; Makino, I. High, But not low, Molecular weight hyaluronan prevents T-cell-mediated liver injury by reducing proinflammatory cytokines in mice. *J. Gastroenterol.* **2004**. [CrossRef]
105. Je, J.Y.; Kim, S.K. Reactive oxygen species scavenging activity of aminoderivatized chitosan with different degree of deacetylation. *Bioorg. Med. Chem.* **2006**. [CrossRef]
106. Coco, G.; Foulsham, W.; Nakao, T.; Yin, J.; Amouzegar, A.; Taketani, Y.; Chauhan, S.K.; Dana, R. Regulatory T cells promote corneal endothelial cell survival following transplantation via interleukin-10. *Am. J. Transplant.* **2020**, *20*, 389–398. [CrossRef] [PubMed]
107. Cheng, K.; Kisaalita, W. Exploring cellular adhesion and differentiation in a micro-/nano-hybrid polymer scaffold. *Biotechnol. Prog.* **2010**, *26*, 838–846. [CrossRef] [PubMed]
108. Ahearne, M.; Fernández-Pérez, J.; Masterton, S.; Madden, P.W.; Bhattacharjee, P. Designing Scaffolds for Corneal Regeneration. *Adv. Funct. Mater.* **2020**, 1908996. [CrossRef]
109. Trujillo-de Santiago, G.; Sharifi, R.; Yue, K.; Sani, E.S.; Kashaf, S.S.; Alvarez, M.M.; Leijten, J.; Khademhosseini, A.; Dana, R.; Annabi, N. Ocular adhesives: Design, chemistry, crosslinking mechanisms, and applications. *Biomaterials* **2019**, *197*, 345–367. [CrossRef]
110. Raghu, P.K.; Bansal, K.K.; Thakor, P.; Bhavana, V.; Madan, J.; Rosenholm, J.M.; Mehra, N.K. Evolution of nanotechnology in delivering drugs to eyes, skin and wounds via topical route. *Pharmaceuticals* **2020**, *13*, 167. [CrossRef]

111. Patra, H.K.; Azharuddin, M.; Islam, M.M.; Papapavlou, G.; Deb, S.; Osterrieth, J.; Zhu, G.H.; Romu, T.; Dhara, A.K.; Jafar M.J.; et al. Rational Nanotoolbox with Theranostic Potential for Medicated Pro-Regenerative Corneal Implants. *Adv. Funct. Mate* **2019**, *29*, 1903760. [CrossRef]
112. Krishna, L.; Dhamodaran, K.; Jayadev, C.; Chatterjee, K.; Shetty, R.; Khora, S.S.; Das, D. Nanostructured scaffold as a determinar of stem cell fate. *Stem Cell Res. Ther.* **2016**, *7*. [CrossRef]
113. Motealleh, A.; Kehr, N.S. Nanocomposite Hydrogels and Their Applications in Tissue Engineering. *Adv. Healthc. Mater.* **201** *6*, 1600938. [CrossRef] [PubMed]
114. Sharma, A.; Taniguchi, J. Review: Emerging strategies for antimicrobial drug delivery to the ocular surface: Implications fc infectious keratitis. *Ocul. Surf.* **2017**, *15*, 670–679. [CrossRef] [PubMed]
115. Bhosale, U.V.; Devi, V.K.; Jain, N. Formulation and optimization of mucoadhesive nanodrug delivery system of acyclovir. *J. Youn Pharm.* **2011**, *3*, 275–283. [CrossRef]
116. Honda, M.; Asai, T.; Oku, N.; Araki, Y.; Tanaka, M.; Ebihara, N. Liposomes and nanotechnology in drug development: Focus o ocular targets. *Int. J. Nanomed.* **2013**, *8*, 495–504. [CrossRef]
117. Vichare, R.; Garner, I.; Paulson, R.J.; Tzekov, R.; Sahiner, N.; Panguluri, S.K.; Mohapatra, S.; Mohapatra, S.S.; Ayyala, R.; Sneec K.B.; et al. Biofabrication of chitosan-based nanomedicines and its potential use for translational ophthalmic applications. *App Sci.* **2020**, *10*, 4189. [CrossRef]
118. Chang, M.-C.; Kuo, Y.-J.; Hung, K.-H.; Peng, C.-L.; Chen, K.-Y.; Yeh, L.-K. Liposomal dexamethasone-moxifloxacin nanopárticl combinations with collagen/gelatin/alginate hydrogel for corneal infection treatment and wound healing. *Biomed. Mater.* **2020**, *1* [CrossRef] [PubMed]
119. Liu, D.; Lian, Y.; Fang, Q.; Liu, L.; Zhang, J.; Li, J. Hyaluronic-acid-modified lipid-polymer hybrid nanoparticles as an efficien ocular delivery platform for moxifloxacin hydrochloride. *Int. J. Biol. Macromol.* **2018**, *116*, 1026–1036. [CrossRef]
120. Gupta, H.; Aqil, M.; Khar, R.K.; Ali, A.; Bhatnagar, A.; Mittal, G. Sparfloxacin-loaded PLGA nanoparticles for sustained ocula drug delivery. *Nanomedicine* **2010**, *6*, 324–333. [CrossRef] [PubMed]
121. Gupta, H.; Aqil, M.; Khar, R.K.; Ali, A.; Bhatnagar, A.; Mittal, G. Biodegradable levofloxacin nanoparticles for sustained ocula drug delivery. *J. Drug Target* **2011**, *19*, 409–417. [CrossRef]
122. Ulyanov, V.A.; Makarova, M.B.; Molchaniuk, N.I.; Ulyanova, N.A.; Skobeeva, V.M.; Chernezhenko, E.A. Effect of colloidal silve nanoparticle solution instillation on the ultrastructure of the corneal epithelium and stroma. *J. Ophthalmol. (Ukraine)* **2017**, 63–69. [CrossRef]
123. Chen, M.; Bao, L.; Zhao, M.; Cao, J.; Zheng, H. Progress in Research on the Role of FGF in the Formation and Treatment of Cornea Neovascularization. *Front. Pharmacol.* **2020**, *11*. [CrossRef]
124. Salama, A.H.; Shamma, R.N. Tri/tetra-block co-polymeric nanocarriers as a potential ocular delivery system of lornoxican In-vitro characterization, and in-vivo estimation of corneal permeation. *Int. J. Pharm.* **2015**, *492*, 28–39. [CrossRef] [PubMed]
125. Üstündağ-Okur, N.; Gökçe, E.H.; Bozbiyik, D.I.; Eğrilmez, S.; Özer, Ö.; Ertan, G. Preparation and in vitro-in vivo evaluation c ofloxacin loaded ophthalmic nano structured lipid carriers modified with chitosan oligosaccharide lactate for the treatment c bacterial keratitis. *Eur. J. Pharm. Sci.* **2014**, *63*, 204–215. [CrossRef] [PubMed]
126. Iriyama, A.; Oba, M.; Ishii, T.; Nishiyama, N.; Kataoka, K.; Tamaki, Y.; Yanagi, Y. Gene transfer using micellar nanovectors inhibit choroidal neovascularization in vivo. *PLoS ONE.* **2011**. [CrossRef]
127. Balakrishnan, S.; Bhat, F.A.; Singh, P.R.; Mukherjee, S.; Elumalai, P.; Das, S.; Patra, C.R.; Arunakaran, J. Gold nanoparticle conjugated quercetin inhibits epithelial–mesenchymal transition, angiogenesis and invasiveness via EGFR/VEGFR-2-mediate pathway in breast cancer. *Cell Prolif.* **2016**, *49*, 678–697. [CrossRef]
128. Cho, W.K.; Kang, S.; Choi, H.; Rho, C.R. Topically administered gold nanoparticles inhibit experimental corneal neovascularizatio in mice. *Cornea* **2015**, *34*, 456–459. [CrossRef]
129. Xu, X.; Sun, L.; Zhou, L.; Cheng, Y.; Cao, F. Functional chitosan oligosaccharide nanomicelles for topical ocular drug delivery c dexamethasone. *Carbohydr. Polym.* **2020**, *227*, 115356. [CrossRef]
130. Zhang, L.; Li, Y.; Zhang, C.; Wang, Y.; Song, C. Pharmacokinetics and tolerance study of intravitreal injection of dexamethason loaded nanoparticles in rabbits. *Int. J. Nanomed.* **2009**, *4*, 175–183. [CrossRef]
131. Ali, H.S.M.; York, P.; Ali, A.M.A.; Blagden, N. Hydrocortisone nanosuspensions for ophthalmic delivery: A comparative stud between microfluidic nanoprecipitation and wet milling. *J. Control. Release* **2011**, *149*, 175–181. [CrossRef] [PubMed]
132. Zhao, X.; Song, W.; Chen, Y.; Liu, S.; Ren, L. Collagen-based materials combined with microRNA for repairing cornea wound and inhibiting scar formation. *Biomater. Sci.* **2019**, *7*, 51–62. [CrossRef] [PubMed]
133. Jafari, S.; Ahmadian, E.; Fard, J.K.; Khosroushahi, A.Y. Biomacromolecule based nanoscaffolds for cell therapy. *J. Drug Deliv. S Technol.* **2017**, *37*, 61–66. [CrossRef]
134. Anitua, E.; Muruzabal, F.; de la Fuente, M.; Merayo, J.; Durán, J.; Orive, G. Plasma Rich in Growth Factors for the Treatment Ocular Surface Diseases. *Curr. Eye Res.* **2016**, *41*, 875–882. [CrossRef] [PubMed]

MDPI
St. Alban-Anlage 66
4052 Basel
Switzerland
Tel. +41 61 683 77 34
Fax +41 61 302 89 18
www.mdpi.com

Micromachines Editorial Office
E-mail: micromachines@mdpi.com
www.mdpi.com/journal/micromachines

www.ingramcontent.com/pod-product-compliance
Lightning Source LLC
LaVergne TN
LVHW070617170726
843515LV00004B/108
* 9 7 8 3 0 3 6 5 3 4 1 6 9 *